Engineering of Submicron Particles

Engineering of Submicron Particles

Fundamental Concepts and Models

Jayanta Chakraborty

Department of Chemical Engineering
Indian Institute of Technology Kharagpur
India

WILEY

This edition first published 2019

Registered Offices
John Wiley & Sons, Inc., 111 River Street, Hoboken, NJ 07030, USA
John Wiley & Sons Ltd, The Atrium, Southern Gate, Chichester, West Sussex, PO19 8SQ, UK

Editorial Office
The Atrium, Southern Gate, Chichester, West Sussex, PO19 8SQ, UK

For details of our global editorial offices, customer services, and more information about Wiley products visit us at www.wiley.com.

Wiley also publishes its books in a variety of electronic formats and by print-on-demand. Some content that appears in standard print versions of this book may not be available in other formats.

Library of Congress Cataloging-in-Publication Data

Names: Chakraborty, Jayanta, 1976- author.
Title: Engineering of submicron particles : fundamental concepts and models / Jayanta Chakraborty, Department of Chemical Engineering, Indian Institute of Technology Kharagpur, India.
Description: Hoboken, NJ, USA : John Wiley & Sons, Inc., [2019] | Includes bibliographical references and index. |
Identifiers: LCCN 2019015637 (print) | LCCN 2019018492 (ebook) | ISBN 9781119296454 (Adobe PDF) | ISBN 9781119296782 (ePub) | ISBN 9781119296461 (hardcover)
Subjects: LCSH: Nanoparticles.
Classification: LCC TA418.78 (ebook) | LCC TA418.78 .C475 2019 (print) | DDC 620.1/15–dc23
LC record available at https://lccn.loc.gov/2019015637

Cover Design: Wiley
Cover Image: Courtesy of Jayanta Chakraborty

Set in 10/12pt Warnock by SPi Global, Pondicherry, India

10 9 8 7 6 5 4 3 2 1

Dedicated to my parents, who had the courage to push us for higher education against many odds.

Contents

Preface

In the process industry, many products and intermediates exist in the form of fine particles. Many next-generation processes, such as colloidal heat transfer fluids for electronic cooling, also involve small particles. However, the ability of the process industry to deal with particulate processes in a quantitative way is limited. The process industry must enhance its capability in the engineering of fine particles.

Many research laboratories also produce and handle submicron particles. In a broader sense, such particulate systems include powders, polymers, colloids or even human populations. While many engineering textbooks and reference books deal with particles of micron scale and above, submicron particles are discussed mostly under very specialized subtopics and a reference book discussing the fundamental concepts of such systems is missing.

Everyday activities in an industrial or academic research laboratory where particulate systems are involved require application of a number of quantitative relations called models. Even experimental facilities use models to relate the raw data with the quantity of interest and often the user is not aware that the outputs are actually from a model. Most models are not straightforward and no single resource is available to provide understanding of frequently used techniques and concepts. New researchers often find themselves at a loss and tend to trust data blindly. This book attempts to resolve this problem by discussing the fundamental theories behind many frequently encountered particulate processes. A large number of diagrams, software, examples, brief experimental demonstrations, and exercises with answers are included and have been carefully planned to provide good learning.

Particulate systems are used by physicists, chemists, mathematicians, and engineers. It is difficult to provide fundamental knowledge to the degree demanded by all. This book is mainly aimed at senior undergraduate or graduate chemical engineering students but provides enough background material in the appendices to be also useful to students from other branches of science and engineering.

Models are used at various levels in particle technology. A set of basic models describe the fundamental process of nucleation, growth, and aggregation of particles. In these models, the rate of nucleation of particles from a medium of given supersaturation, the rate of increase of size of a particle of given size under a set of environmental conditions, and the rate of aggregation of given pairs are provided.

Classical nucleation theory is discussed at length in this text. Other nucleation mechanisms, e.g. the organizer mechanism, are also introduced. For growth, the classical growth models such as diffusion controlled and surface nucleation controlled growth are discussed, along with newer models like connected net analysis. Aggregation models and inter-particle potentials are discussed with a brief but useful prelude on inter-molecular and surface forces.

The basic models alone cannot describe the dynamics of an engineering system containing a large number of particles of varying attributes. For this a number balance equation (population balance) is needed. In this book the emphasis is on formulating the number balance equation (the population balance model) for a given system. Analytical and numerical solutions of population balance models are also discussed briefly. Software with open code is provided for the solution of a population balance model through discretization.

To my knowledge no book serves such a diverse yet unified purpose. This book has been in my mind throughout my career over the past decade, during which I made my journey from an experimental laboratory to two theoretical laboratories and then back to experiments. This book contains useful insights which I acquired over time.

This book is heavily indebted to several books and monographs which helped me in assimilating the content. I kept close to the flow of ideas and concepts of the parent books whenever I felt that was best for the reader. I acknowledge major contributions from the following books and monographs:

- *Foundations of Colloid Science* by R. J. Hunter and *Kinetics of Precipitation* by A. E. Nielsen for the nucleation and growth chapters.
- The chapter on inter-molecular and inter-particle force has ideas and contents from *Intermolecular and Surface Forces* by Jacob Israelachvili.
- The stability chapter is heavily indebted to Paul C. Hiementz (*Principles of Colloid and Surface Chemistry*).
- The particulate system modelling section is indebted to *Population Balances* by D. Ramkrishna and *Theory of Particulate Processes* by A. D. Randolph and M. A. Larson.
- Much of the book is also influenced by the lecture notes circulated during my graduate course on modelling at the Indian Institute of Science, Bangalore by Prof. K. S. Gandhi and Prof. Sanjeev Kumar.

Apart from these major resources there are many other books and monographs that helped me to understand, assimilate, and express the ideas. I also acknowledge help from students at IIT Kharagpur who took this course (Fundamentals of Particle Technology, CH60026), asked critical questions, and helped me write this book. I hope this book will be useful to the others. Of course there are multiple errors and omissions which I'm eager to hear from the readers and correct in a future edition.

Jayanta Chakraborty
IIT Kharagpur
Autumn 2018

About the Companion Website

To access supplementary materials for this book please use the download link shown below:

http://booksupport.wiley.com

Please enter the book title, author name or ISBN to access this material.

Here you will find valuable material designed to enhance your learning, including:

- Demo Videos
- Source Codes

1

Nucleation

Nucleation means the creation of new particles, i.e. the creation of a new phase and associated interface. Hence, in order to understand the nucleation process, we need to learn a few key ideas from physical chemistry. The energy of the interface is vital to the nucleation process, which in turn controls the nucleation rate. Hence, our discussion will involve both thermodynamics and kinetics: the thermodynamics of the interface will provide the magnitude of the driving force and the kinetics will provide the rate of nucleation.

1.1 Thermodynamics of Interfaces

An interface is a surface where one phase ends and another starts. It is a narrow region often in the order of a few angstroms where the properties change from that of one phase to another. For a liquid–vapour interface, the density of the medium undergoes an abrupt change. For a liquid–liquid interface, two dissimilar atoms are in contact at the interface.

1.1.1 The Interface is a Surface of High Energy

An interface is known to contain higher energy than the bulk phase. This difference in energy is key to phenomena relating to many important technical problems. The excess energy of the interface, or interfacial energy, is due to the difference between the energies of atoms on surface and in the bulk. The difference may be due to the change in density between the two phases or to the difference in chemical nature.

Let us consider the former as an example. For a liquid–vapour interface the liquid is in contact with its vapour. Hence, although similar atoms/molecules are present on both sides of the interface, the densities are very different. This leads to a different coordination number of atoms in bulk versus atoms on the surface. Next we show how this leads to interfacial energy.

Engineering of Submicron Particles: Fundamental Concepts and Models, First Edition. Jayanta Chakraborty.

Companion website: http://booksupport.wiley.com

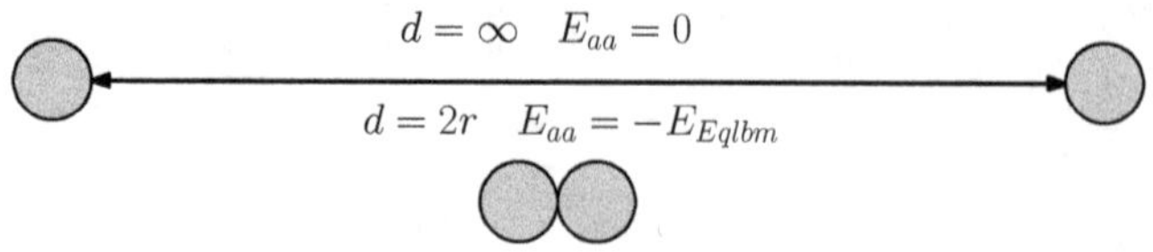

Figure 1.1 Distance-dependant interaction potential. The negative sign indicates energy release.

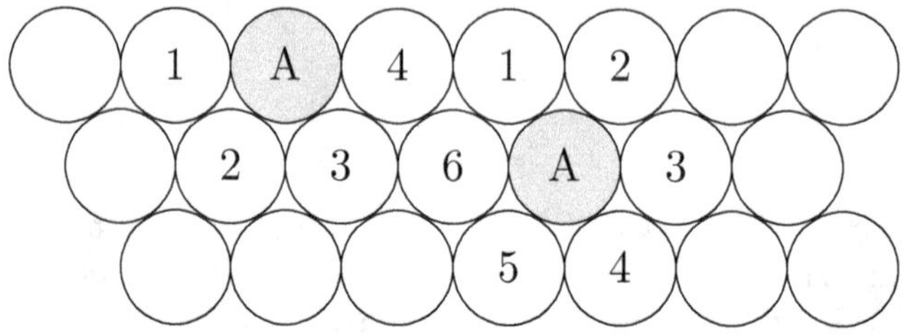

Figure 1.2 Difference in coordination number between surface and bulk atoms. In this case, the surface atom coordinates with only four other atoms while the bulk atom coordinates with six.

The energy of interaction of two isolated atoms at infinite separation is zero. If they are brought closer together, they start to interact. At a specific distance, $2r$, contact between the two atoms, the energy of interaction, becomes a minimum.

Let us denote the amount of energy *released* by bringing a pair of atoms of same chemical species to this distance by E_{AA}. Hence, the energy per atom for constructing a pair is $E_{AA}/2$. What is the energy released per atom where all atoms are jam-packed, as shown in Figure 1.2? A pair of simplifications will be useful in analyzing the case:

- only the nearest neighbours of an atom can impart some force on an atom
- interaction is pairwise additive.

With these assumptions, an atom in the bulk (see Figure 1.2) will release energy corresponding to the pairs A-1, A-2, … A-6 as shown in the figure (the coordination number is 6). If we denote the coordination number in bulk by z_b, the total energy *released* per atom will be

$$-\frac{z_b E_{AA}}{2}.$$

Now, the coordination number is different for the surface. Hence, the energy *released* per atom for the surface will be

$$-\frac{z_s E_{AA}}{2}$$

where the coordination number for the surface is z_s. Because $z_b > z_s$, more energy is *released* for the atoms in the bulk than on the surface. Hence, the surface atoms *retain* more energy. Hence the system that has more interface, has more energy. In other words, interface contains energy.

1.1.2 The Interface is a Surface Under Tension

Will the surface resist its extension? It should. More surface will require more atoms to join the surface, leaving the bulk, and hence it goes energy uphill. Hence, interfaces normally act like a stretched membrane.

The extra interface will require extra energy, which will be supplied by external work. If we denote the energy needed for the creation of a differential amount of surface δA by δW, the following proportionality can be written:

$$\delta W \propto \delta A.$$

Inserting the constant of proportionality:

$$\delta W = \gamma \delta A.$$

It is clear that the constant γ is the surface energy per unit area and hence is interpreted as the specific surface energy. If work is done by a constant external force F_s to increase the area and the increase in area can be written as $\delta A = l\delta x$, the above equation becomes

$$F_s \delta x = \gamma l \delta x$$

or

$$F_s / l = \gamma,$$

which leads to the popular interpretation of specific surface energy as 'surface tension' with unit force/length.

1.1.3 Pressure Drop Across Curved Interfaces

The higher energy of the interface leads to difference in pressure across a *curved* interface [1]. Let us consider a small area, as shown in the Figure 1.3, and perturb the surface by varying the pressure differentially: the surface expands differentially in response to the differential increase in pressure. The increase in area is

$$(x + dx)(y + dy) - xy = xdy + ydx.$$

This additional area will require additional surface energy, which is

$$\gamma(xdy + ydx).$$

This much energy must be supplied by working against a difference in pressure (ΔP) between the two sides of the curved interface. If the inside pressure during the perturbation changes only differentially, the pressure difference across the

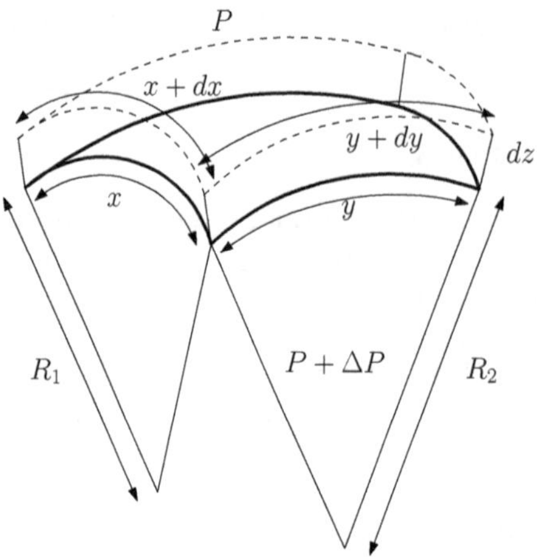

Figure 1.3 The geometry of a curved interface: derivation of the Young–Laplace equation

interface remains ΔP even after the perturbation. Hence the PV work is given by $\Delta P \Delta V$. The increase in volume, ΔV, in this case is given by $xydz$ and hence the energy balance can be written as

$$\gamma(xdy + ydx) = \Delta Pxydz$$

or

$$\frac{\Delta P}{\gamma} = \frac{xdy + ydx}{xydz} = \frac{1}{y}\frac{dy}{dz} + \frac{1}{x}\frac{dx}{dz}.$$

Using the property of similar triangles, we can write

$$\frac{x}{R_1} = \frac{x + dx}{R_1 + dz}$$

and

$$\frac{y}{R_2} = \frac{y + dy}{R_2 + dz}.$$

These two expressions lead to

$$\frac{1}{R_1} = \frac{1}{x}\frac{dx}{dz}$$

and

$$\frac{1}{R_2} = \frac{1}{y}\frac{dy}{dz}.$$

Hence, the above equation can be written as

$$\frac{\Delta P}{\gamma} = \frac{1}{R_1} + \frac{1}{R_2}. \tag{1.1}$$

This is known as the Young–Laplace equation and gives the pressure difference across a curved interface as a function of its curvature.

Example 1.1 What is the pressure inside a small water droplet of radius 1 μm and one with radius 1 nm? The surface tension of water is 75 mN/m.

Solution: Because the drop is spherical, both the radii are equal in this case. Hence, the Young–Laplace equation reduces to

$$\frac{\Delta P}{\gamma} = \frac{2}{R}.$$

Hence for a 1 μm drop:

$$\Delta P = \frac{2 \times 75 \times 10^{-3}}{1 \times 10^{-6}} = 0.15 \times 10^{6}\ \text{Pa} = 1.5\ \text{atm}.$$

If the drop size is 1 nm,

$$\Delta P = \frac{2 \times 75 \times 10^{-3}}{1 \times 10^{-9}} = 150 \times 10^{6}\ \text{Pa} = 1500\ \text{atm}.$$

It can be seen that for the first case the pressure difference is merely 1.5 atm whereas for the later it is huge: 1500 atm. Usually, nuclei are very small, of the order of nanometres, and hence they experience huge pressure due to the curved interface. □

Example 1.2 What is the pressure inside a small soap bubble of radius 1 cm?

Solution: Because the soap bubble has two interfaces, the Young–Laplace equation should be written for both interfaces. Denoting P_i as inside pressure, P_f as film pressure, and P_o as outside pressure, and applying the Young–Laplace equation for both interfaces,

$$P_i - P_f = \frac{2\gamma}{R}$$

$$P_f - P_o = \frac{2\gamma}{R + \delta R}$$

$$\therefore P_i - P_o = \frac{2\gamma}{R} + \frac{2\gamma}{R + \delta R}.$$

Neglecting the film thickness δR:

$$\therefore P_i - P_o = \frac{4\gamma}{R} = 30\ \text{Pa}.$$

Note that we have used the surface tension value of water instead of the surfactant solution in order to obtain an approximate value. The true surface tension of the surfactant solution is dependent on the nature and concentration of the surfactant and should be used for an accurate value. □

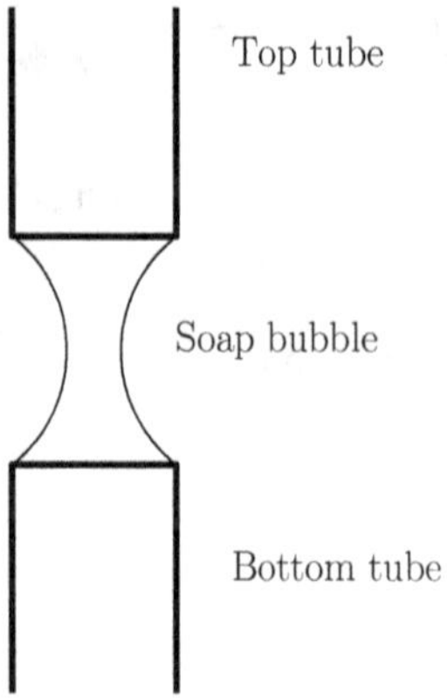

Figure 1.4 Surface of zero net curvature.

Example 1.3 A soap bubble can be stabilized on a pair of open tubes, as shown in the Figure 1.4. The pressure is 1 atm on both sides of the bubble yet it has a curved interface. Does it contradict the Young–Laplace equation?

Solution: For any curved surface, the curvatures in two mutually perpendicular directions are related by the Young–Laplace equation. For the above case the pressure difference is zero and hence the Young–Laplace equation reduces to

$$\frac{1}{R_1} + \frac{1}{R_2} = 0.$$

For the surface shown in Figure 1.4, one curvature is convex while the other is concave to the observer from either side. Hence, one curvature is positive and the other is negative, and their values should be such that the Young–Laplace equation is exactly satisfied. Such surfaces are called surfaces of zero average curvature. □

1.1.3.1 Capillary Rise

One of the interesting manifestations of the pressure difference across a curved interface is capillary rise. Figure 1.5 shows a capillary tube and the liquid meniscus. The liquid–vapour interface is a curved surface in the capillary because of a phenomena known as the contact angle. Every liquid–solid–vapour interface maintains a definite angle of contact, as shown in the figure, and hence the interface cannot remain flat. If the tube has a larger diameter the interface is nearly flat except for the edges and if the tube diameter is small (a capillary tube) the interface becomes nearly spherical.

In our previous example, we dealt with a liquid drop and showed that the pressure is higher inside the drop. For the liquid meniscus in a capillary, which side has the higher pressure? It is determined by the sign of the curvature. For a spherical surface this boils down to the simple rule that the side which contains

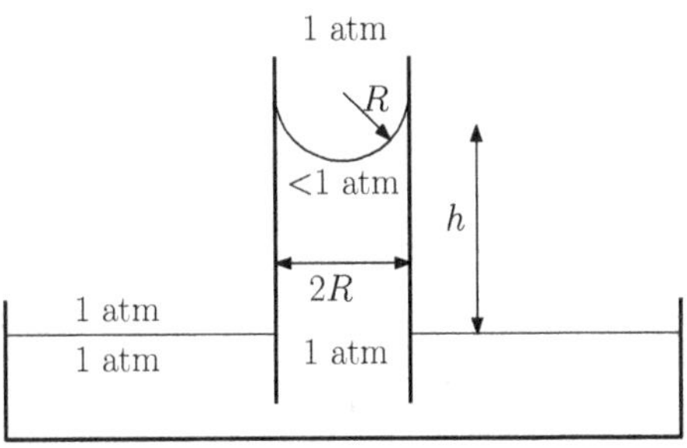

Figure 1.5 The contact angle and curved interface in a capillary tube: the reason for the elevated liquid level in a capillary tube.

the centre has higher pressure. Hence, for the case of a liquid meniscus in a capillary, the pressure in the gas phase is higher.

If the meniscus to co-inside with the liquid level, the pressure difference cannot be accounted for because the pressure at that level is 1 atm in both the liquid and the gas. But if the liquid level is raised, the pressure on the gas side is 1 atm but the pressure in the liquid side at that raised level is less than 1 atm. It reaches 1 atm at the surface of the liquid pool by addition of the gravity head of the liquid column. A force balance yields:

$$h\rho g = \frac{2\gamma}{R},$$

where h is the capillary rise.

Example 1.4 What will be the elevation of the liquid level in a capillary whose radius is 100 μm? The surface tension of water is 75 mN/m.

Solution:

$$h = \frac{2\gamma}{R\rho g} - \frac{2\times 75\times 10^{-3}}{100\times 10^{-6}\times 10^{3}\times 10} = 15\times 10^{-2}\text{m} = 15\text{cm}$$

□

In the above example, it can be seen that significant capillary rise is possible. Moisture is often drawn into narrow inter-particle spaces through capillary action and it is difficult to drive away such trapped moisture.

1.1.4 Vapour–Liquid Equilibrium Across Curved Interfaces

One of the reasons for our interest in the pressure difference in a small drop is that it alters the chemical potential of the drop. The chemical potential of a phase is a function of pressure, temperature, and composition, and hence excess pressure inside a drop alters its chemical potential. In other words, for a pure substance under isothermal conditions, droplets of different sizes will have different chemical potentials. The quantitive description of size-dependent chemical potential is given by Kelvin's equation.

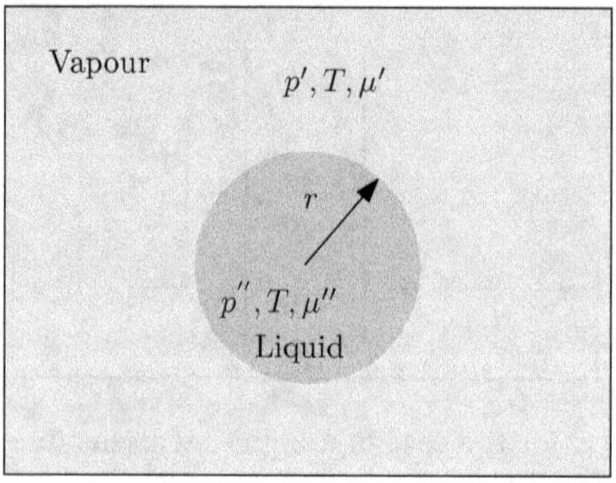

Figure 1.6 Vapour–liquid equilibrium in an isothermal system.

Let us consider a closed isothermal vessel containing a pure vapour maintained at temperature T and pressure p', as shown in Figure 1.6. The chemical potential of the vapour phase is μ'. Now we ask the following question:

What is the Radius of a Drop Which is in Equilibrium with this Vapour?

Let us denote the chemical potential and pressure of the condensed phase by μ'' and p'', respectively. The pressure inside the drop is given by the Young–Laplace equation:

$$p'' = p' + \frac{2\gamma}{r}.$$

Notice that we have used the condition of mechanical equilibrium, but the condition for chemical equilibrium remained unused. Hence the desired relation may be obtained by using chemical equilibrium. Chemical equilibrium enforces

$$\mu'' = \mu'$$

or

$$d\mu'' = d\mu'. \tag{1.2}$$

Now, the relation between chemical potential (partial molar Gibbs energy) and other thermodynamic variables is given by the Gibbs–Duhem equation:

$$\left(\frac{\partial g}{\partial p}\right)_T dp + \left(\frac{\partial g}{\partial T}\right)_p dT - d\mu = 0.$$

Substituting

$$\left(\frac{\partial g}{\partial p}\right)_T = v$$

and restricting to isothermal cases, we obtain

$$v'dp' = v''dp''.$$

Now differentiating the Young–Laplace equation

$$dp'' = dp' + d\left(\frac{2\gamma}{r}\right) \tag{1.3}$$

and substituting dp'' from the above we get

$$\left(\frac{v'dp'}{v''} - dp'\right) = d\left(\frac{2\gamma}{r}\right)$$

$$\frac{v' - v''}{v''}dp' = d\left(\frac{2\gamma}{r}\right), \tag{1.4}$$

which is the basic form of Kelvin's equation. A more usable form may be obtained by assuming $v' >> v''$ and the vapour phase to be an ideal gas. These assumptions lead to

$$\frac{v'}{v''}dp' \approx d\left(\frac{2\gamma}{r}\right)$$

or

$$\frac{RT}{v''p'}dp' \approx d\left(\frac{2\gamma}{r}\right).$$

This equation can be integrated and the constant of integration can be evaluated by considering the limit where the interface is flat, i.e. its radius of curvature is infinity:

$$r \to \infty \Rightarrow \frac{2\gamma}{r} \to 0.$$

The pressure difference across a flat interface is zero and the equilibrium pressure corresponding to such a flat interface is generally available and reported as the 'saturation pressure'. Let us denote it by p_0. We can also safely assume that the molar volume of the liquid is nearly constant in the range of pressure considered. Hence, the above expression can be integrated as:

$$\int_{p_0}^{p'} \frac{dp'}{p'} \approx \frac{v''}{RT} \int_0^{2\gamma/r} d\left(\frac{2\gamma}{r}\right)$$

$$\ln\frac{p'}{p_0} \approx \frac{2\gamma v''}{RTr}. \tag{1.5}$$

This expression provides the equilibrium radius of a drop as a function of the prevailing pressure and is known as Kelvin's equation.

Example 1.5 Water vapour at 100 °C is initially stored at 101 kPa. If a 1 nm water droplet is to remain at equilibrium in this vessel at the same temperature, how much supersaturation should be created? The density of saturated water can be taken as 1000 kg/m^3.

Solution: The approximate molar volume of liquid water is 18 ml/mol = 18 × 10^{-6} m^3/mol. Taking the surface tension of water to be 75 × 10^{-3} N/m,

$$\ln\frac{p'}{p_0} = \frac{2\times 75\times 10^{-3}\times 18\times 10^{-6}}{8.314\times 373\times 1\times 10^{-9}} = 0.870$$

$$\therefore \frac{p'}{p_0} = 2.4.$$

□

It is clear from the above example that considerable supersaturation is needed to keep the nucleus at equilibrium.

Example 1.6 What is the vapour pressure inside a small bubble in a liquid? Calculate the vapour pressure for a 5 nm vapour bubble at 373 K.

Solution: We note that the development of the final form of Kelvin's equation as given by eqn 1.5 is tied to the fact that the drop was the condensed phase. To treat a situation where we form a bubble, we need to go back a few steps. We retain the notation that the condensed phase is indicated by $''$. Using the set of fundamental equations

$$v'dp' = v''dp''$$

and

$$dp' = dp'' + d\left(\frac{2\gamma}{r}\right)$$

and eliminating dp'',

$$\left(\frac{v''-v'}{v''}\right)dp' = d\left(\frac{2\gamma}{r}\right).$$

Now, using the pair of assumptions we used before,

$$-\frac{RT}{v''}\frac{dp'}{p'} = d\left(\frac{2\gamma}{r}\right).$$

Integrating between the limit between the flat and curved interfaces:

$$\ln\frac{p'}{p_0} \approx -\frac{2\gamma v''}{RTr}.$$

Substituting numerical values we get $p'/p_0 = 0.84$. □

1.1.4.1 Thomson Equation

In the above cases, isothermal conditions were maintained and the pressure was varied. The equilibrium temperature of a drop of given size can also be obtained for isobaric conditions. This is given by Thomson's equation and is obtained as follows.

Again, the chemical equilibrium is used along with the Young–Laplace equation, but in this case we set $dp' = 0$ in the Gibbs–Duhem equation:

$$s'dT = s''dT - v''dp''$$

or

$$(s' - s'')dT = -v''dp''.$$

Imposing isobaric conditions in the vapour phase ($dp' = 0$) in the Young–Laplace equation (eqn 1.3) and substituting for dp'' in the above equation, we get

$$(s' - s'')dT = -v''d\left(\frac{2\gamma}{r}\right).$$

Now, for an isobaric process:

$$(s' - s'') = \frac{\Delta h_{vap}}{T}.$$

Substituting the above relation:

$$\frac{\Delta h_{vap}}{T}dT = -v''d\left(\frac{2\gamma}{r}\right).$$

Integrating between a flat and a curved interface of radius r:

$$\Delta h_{vap}\int_{T_0}^{T}\frac{dT}{T} = -2v''\gamma\int_{\infty}^{r} d\left(\frac{1}{r}\right)$$

or

$$\Delta h_{vap}\ln\left(\frac{T}{T_0}\right) = -\left(\frac{2v''\gamma}{r}\right)$$

or

$$\ln\left(\frac{T}{T_0}\right) = -\left(\frac{2v''\gamma}{\Delta h_{vap}r}\right), \tag{1.6}$$

which is known as Thomson's equation.

Example 1.7 Water vapour is stored in a closed vessel at 101 kPa pressure. If a 1 nm water droplet is to remain at equilibrium in this vessel at this pressure, how much cooling should be provided under isobaric conditions? The density of saturated water can be taken as 1000 kg/m^3 and may be considered constant. The enthalpy of vaporization is 2257 kJ/kg.

Solution: The molar volume of liquid water is 18 ml/mol = 18×10^{-6} m^3/mol. Taking the surface tension of water to be 75×10^{-3} N/m, $\Delta H_{vap} = 2257 \times 18/1000$ kJ/mol $= 2257 \times 18$ J/mol.

$$\ln \frac{T}{T_0} = -\frac{2 \times 75 \times 10^{-3} \times 18 \times 10^{-6}}{2257 \times 18 \times 10^{-9}} = -0.0664$$

$$\therefore \frac{T}{T_0} = 0.94.$$

Although this appears small, it signifies a large amount of supercooling. In °C, the vapour needs to be cooled to 76 °C! Again, it is clear that huge supercooling is needed to maintain such small drops at equilibrium. □

1.1.5 Stability of the Equilibrium

The equilibrium described by Kelvin/Thomson's equation is unstable. Let us consider the isothermal case described by the Kelvin equation (eqn 1.5). It can be seen from the equation that the equilibrium pressure is inversely related to drop size. As drop size increases, the equilibrium pressure decreases and vice versa. Now, imagine a drop of size r existing in equilibrium at a pressure p. What will happen if this drop grows slightly? The equilibrium pressure for this slightly bigger drop is less than p. Hence, for this drop a pressure p means a supersaturated environment so it will grow.

Now, tiny fluctuations are always present in any system. If an equilibrated drop increases in size by tiny fluctuations, its environment becomes supersaturated and it grows. The more it grows, the more supersaturated the environment becomes and the more rapid the growth.

Now, if it shrinks by a small amount, its equilibrium pressure is higher. Hence, to this drop, the existing pressure is less than its equilibrium pressure so it must evaporate. The more it evaporates, the more unsaturated its environment becomes and the more evaporation happens.

In both cases, tiny fluctuations dislodge the system from its equilibrium state and promote rapid growth or shrinkage of the drop. Hence, the equilibrium described by Kelvin/Thomson's equation is not a stable equilibrium.

Example 1.8 Saturated water vapour (at 100 °C and 1 atm) is stored in a closed container. The container also contains liquid drops ranging in size from a few nanometres to micrometres. Will the system remain as it is or will it evolve with time?

Solution: At 100 °C and 1 atm pressure, the saturation pressure for a micrometre-sized drop is 1.0024 atm whereas that for the nanometre-sized drops is 2.4 atm. For both types of drop the environment is unsaturated (1.000 atm). Hence both types of drop will evaporate. The nanometre-sized drops will evaporate more quickly than the micrometre-sized drops.

Where will the evaporated material go? At constant temperature and volume, it will increase the pressure inside the container. When the pressure reaches 1.0024 atm the larger drops stop evaporating. However, the smaller drops continue to evaporate because the environment is still unsaturated for them.

Eventually, the pressure will exceed 1.0024 atm. At this point the larger drops will start to grow. Usually smaller drops evaporate very quickly and disappear from the system, and the material is deposited on the larger drops as explained above. This process is called Ostwald ripening. □

1.2 Homogeneous Nucleation

If a vapour is maintained at a temperature higher than its saturation temperature at the prevailing pressure, the vapour remains vapour indefinitely. The vapour remains vapour even if the temperature is *at* the saturation temperature. However, if the temperature is 'sufficiently' lower than the saturation temperature, particles of liquid are formed in the vapour. A large number of small liquid particles, called nuclei, are formed within a short period. This process is called nucleation. The difference between the temperature of the vapour and its saturation temperature is called supersaturation. The higher the supersaturation, the higher the rate of nucleation. No aid or any additional material is needed for formation of the new phase and nuclei form throughout the vapour phase uniformly or homogeneously. This phenomenon is called homogeneous nucleation.

1.2.1 Energetics of Homogeneous Nucleation

Let us consider a closed vessel of constant volume V_T containing n_i moles of a vapour at constant temperature T. The initial pressure in the vessel is p'. **We want to calculate the change in Helmoltz free energy when a liquid droplet of radius r is formed by condensing a part of the vapour.** This process is shown in Figure 1.7.

Since the formation of a droplet will reduce the mass of vapour in the vessel, the pressure will change after the formation of the drop, but the total volume and temperature will remain constant. We shall analyze the energetics of the system using Helmholtz free energy [2]. However, other forms of free energy can also be used. For a single-component system, the Helmholtz free energy of n mol of a substance (single phase) is given by

$$F = -pV + \mu n.$$

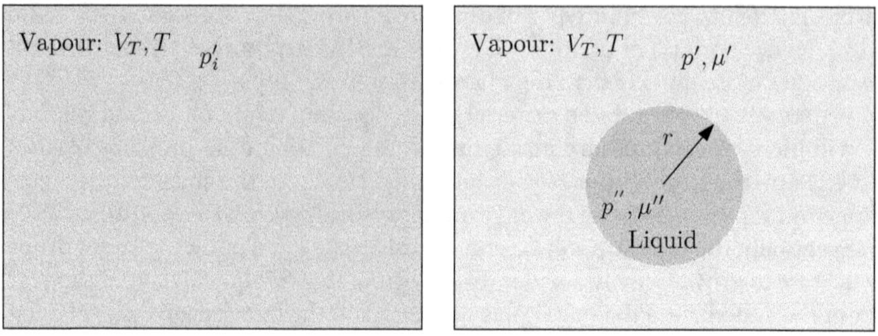

Figure 1.7 Formation of a liquid nucleus in a supersaturated vapour.

Initially, the system pressure is maintained at p'_i and contains only the vapour. Hence the total free energy of the system will be

$$F_i = -p'_i V_T + \mu'(p'_i) n_T.$$

Now, if some of the vapour condenses to form a drop of radius r, the pressure in the vapour phase becomes p'_f and the free energy of the system changes to

$$F_f = (-p'_f V' + \mu'(p'_f)n') + (-p'' V'' + \mu''(p'')n'') + \gamma A,$$

where the primed quantities refer to those of the vapour phase and the double primed quantities refer to those of the condensed phase. A is the surface area of the drop. Because the surface is created, a surface free energy contribution is needed in free energy accounting. An useful simplification is to assume that the container is large enough so that creation of a tiny drop does not change the pressure substantially and hence $p'_f = p'_i = p'$. With this simplification, the free energy change for the process becomes

$$F_f - F_i = (-p'V' + \mu'(p')n') + (-p''V'' + \mu''(p'')n'') + \gamma A - (-p'V_T + \mu'(p')n_T).$$

Recognizing that

$$V' + V'' = V_T \qquad \text{and} \qquad n' + n'' = n_T$$

we obtain

$$F_f - F_i \equiv \Delta F = -V''(p'' - p') + n''[\mu''(p'') - \mu'(p')] + \gamma A.$$

Now, the chemical potential difference can be found by a similar analysis as shown for Kelvin's equation and is given by

$$d\mu''(p'') - d\mu'(p') = v''dp'' - v'dp' = v''\left[dp' + d\left(\frac{2\gamma}{r}\right)\right] - v'dp'.$$

Using the same line of arguments and simplifications as before:

$$\mu''(p'') - \mu'(p') = -RT\ln\left(\frac{p'}{p_0}\right) + v''\left(\frac{2\gamma}{r}\right).$$

Substituting the difference in pressure from the Young–Laplace equation and the chemical potential difference from the above expression, we get the free energy change for nucleation as

$$\Delta F = -n''RT\ln\left(\frac{p'}{p_0}\right) + \gamma A. \tag{1.7}$$

This expression can be written in terms of drop size (radius) as

$$\Delta F = -\left(\frac{4\pi r^3}{3v''}\right)RT\ln\left(\frac{p'}{p_0}\right) + 4\pi\gamma r^2. \tag{1.8}$$

The change in free energy has two parts. The first part is negative and represents the volume free energy change due to condensation. The second part is the surface free energy, which is always positive. This portion of free energy resists the formation of new phase.

The plot of free energy change as a function of particle size (number of molecules) is shown in Figure 1.8. It can be seen that it goes through a maximum. The maximum occurs because of the interplay between the negative volume free energy contribution and a positive surface free energy contribution. It is readily shown that the maximum occurs at

$$r^* = \frac{2\gamma v''}{RT\ln(p'/p_0)} \tag{1.9}$$

and the maximum free energy penalty is given by:

$$\Delta F^* = \frac{4}{3}\pi(r^*)^2\gamma. \tag{1.10}$$

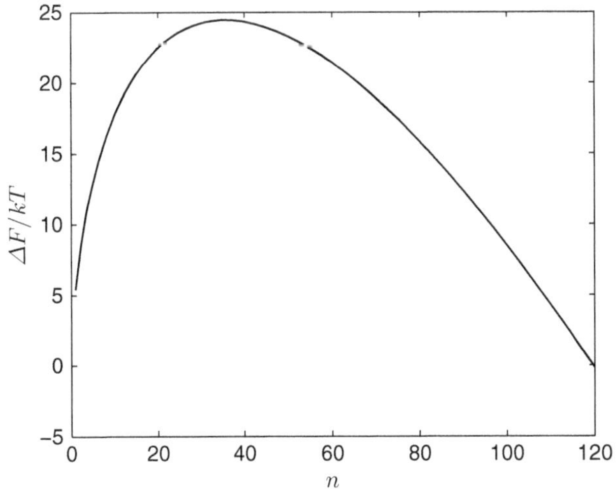

Figure 1.8 $\Delta F/kT$ as a function of n for $\phi/kT = 1.38$ and $\psi/kT = 6.8$.

1.2.1.1 Energetics in Terms of Number of Units

n'' in eqn 1.7 denotes the 'moles' of units (atoms/molecules) in the nucleus. A more helpful quantity, however, is the absolute number of units [3] in a nucleus (n), which is given by $n = Nn''$. Both the volume and the surface area of the nucleus may be expressed in terms of n. The volume is simply nv_m, where v_m is the volume of one unit and the surface area of a nucleus containing n units is given by

$$A_n = (4\pi)\left(\frac{3nv_m}{4\pi}\right)^{2/3} = \sqrt[3]{36\pi v_m^2}\quad n^{2/3}.$$

Now, eqn 1.7 can be written in terms of n as

$$\Delta F = -nkT\ln\left(\frac{p'}{p_0}\right) + \gamma\sqrt[3]{36\pi v_m^2}\quad n^{2/3},$$

where $k = R/N$ is the Boltzmann constant. The constant $(36\pi v_m^2)^{1/3}$ is the area corresponding to a single unit and hence $(36\pi v_m^2)^{1/3}\gamma$ can be considered as the 'surface energy per unit'. Note that this assignment is purely notional. Denoting $(36\pi v_m^2)^{1/3}\gamma$ as ψ,

$$\Delta F = -nkT\ln\left(\frac{p'}{p_0}\right) + \psi n^{2/3}.$$

We can also define volume free energy per unit, i.e. $kTln\left(\frac{p'}{p_0}\right)$, as ϕ. The above equation then becomes

$$\Delta F = -n\phi + \psi n^{2/3}.$$

The criticality observed in Figure 1.8 can be obtained as

$$n^* = \left(\frac{2\psi}{3\phi}\right)^3$$

and

$$\Delta F^* = -n^*\phi + \psi\left(\frac{2\psi}{3\phi}\right)^2 = -n^*\phi + \frac{3\phi}{2}\left(\frac{2\psi}{3\phi}\right)^3 = \frac{\phi n^*}{2}.$$

Example 1.9 Determine the size of the critical nucleus and the magnitude of the critical free energy change if a pressure supersaturation of 4 is maintained in water–water vapour system at 373 K.

Solution: The volume free energy per molecule is

$$\phi = kT\ln\left(\frac{p'}{p_0}\right) = 1.38kT.$$

The surface free energy per molecule is

$$\psi = (36\pi v_m^2)^{1/3}\gamma = \sqrt[3]{36\pi(3\times10^{-29})^2}\times 75\times10^{-3}\text{J}$$
$$= 350\times10^{-22}\text{J}.$$

In terms of kT,

$$\psi = \frac{350\times10^{-22}}{5.15\times10^{-21}}kT = 6.8kT,$$

which gives $n^* = 35$.

The variation of n^* and the activation energy barrier ΔF with supersaturation can be tabulated as follows:

p/p_0	n^*	$\Delta F/kT$
1.01	1×10^9	6.9×10^8
1.50	1397	9.63×10^2
4.0	34	2.35×10^1
10	8	5.52
20	3.5	2.4

It can be seen that the free energy barrier becomes comparable with kT near a supersaturation ratio of 20. □

It should be noted that for particles smaller than the critical size, the free energy increases when growth occurs. On the other hand, free energy decreases due to growth for particles larger than the critical size. Hence, (cf. Figure 1.8) particles above the critical size grow spontaneously. In the following discussions, particles above the critical size will be called nuclei. We will refer to the particles of the critical size as critical nuclei. The particles smaller than the critical size will be called embryos. As we will see, some of the embryos grow, cross the energy hill, and become nuclei then finally grow into large particles.

1.2.2 Kinetics of Homogeneous Nucleation

In the previous section formation of a single nucleus/embryo was discussed. In reality, a large number of such particles are formed in a given volume. The collection of such particles can be visualized in two ways:

1) A large number of embryos/nuclei is formed but they stay in very close proximity in the form of a cluster.
2) The embryos/nuclei are scattered uniformly over the entire available volume.

In both cases, the free energy change can be calculated from eqn 1.8 and will be the same. But do they have same free energy? Since one of the systems (the distributed one) has more entropy than the other and free energy has an entropic part (which we have ignored altogether so far because we considered a single particle), the free energy cannot be same. The entropic contribution of free energy is required when considering a large number of particles.

1.2.2.1 Concentration of Embryos/Nuclei

If we imagine the particles as dissolved chemical species, the entropic contribution to free energy is given by $RT\ln(x_n)$ for one *mole* of this 'species'. Here x_n is the mol fraction of embryos/nuclei having n units. Hence, for a single embryo/nucleus, the entropic contribution will be $kT\ln(x_n)$ and the free energy change per embryo/nucleus is given by

$$\Delta F = -n\phi + \psi n^{2/3} + kT\ln(x_n). \tag{1.11}$$

From a purely energetic perspective, only one size is in equilibrium at a given supersaturation. However, other sizes also exist in the system at equilibrium when the entropic term is included, but their equilibrium concentration is dependent on their energy penalty. Particles which have a higher energy penalty will have lower concentration.

Now, setting $\Delta F = 0$ (at equilibrium) in eqn 1.11, we can obtain the equilibrium concentration of these species:

$$kT\ln(x_n^e) = n\phi - \psi n^{2/3}$$

or

$$x_n^e = \exp[(n\phi - \psi n^{2/3})/kT].$$

This equation relates the equilibrium mol fraction of embryos/nuclei to measurable quantities, namely, the supersaturation and surface energy.

The concentration of embryos in the number/unit volume is a more useful measure of concentration than the mol fraction. To relate this quantity with the mol fraction, let us consider one mol of vapour and associated embryos and nuclei. The total number of embryos/nuclei considering all sizes are usually much smaller than the total number of vapour molecules present and hence one mol of such a mixed substance will approximately occupy the molar volume of the vapour, v. Hence, the equilibrium number concentration of embryos containing n units can be written as

$$C_n^e = \frac{x_n^e N}{v} = \frac{1}{v_m}\exp[(n\phi - \psi n^{2/3})/kT]. \tag{1.12}$$

In the above equation v/N is denoted as v_m, the notional volume per molecule.

Example 1.10 Compute the equilibrium concentrations of embryos/nuclei of various sizes.

Solution: Using the same parameter values as in the previous examples, the following table can be prepared:

n	$(n\phi - n^{2/3}\psi)/kT$	C_n^e (No./m³)
2	−8.03	1.1×10^{25}
5	−13.00	7.6×10^{22}
10	−17.80	6.4×10^{20}
35	−24.45	8.0×10^{17}
50	−23.30	2.6×10^{18}
100	−8.50	6.8×10^{24}
120	+0.164	4.0×10^{28}

□

It can be seen that the concentration is a minimum for the critical size. This is expected because this size has the highest free energy penalty. It can also be seen that for large n, the argument of the exponential becomes positive and the *equilibrium* number concentration becomes very high. This is because the equilibrium assumption is not valid for such large sizes. Particles at this size grow at a very fast rate and do not reach any equilibrium.

1.2.2.2 Chain of Reactions Towards Formation of Nuclei

In a nucleating system embryos of all sizes are present. The embryos which are much larger than the critical size (called nuclei) grow rapidly and produce particles. Embryos of various sizes may form either by attachment–detachment of individual units or due to interactions among embryos themselves. Because embryo concentration is much less than that of vapour molecules, there are far fewer embryo–embryo interactions than embryo–primary unit interactions. Hence, the system mostly evolves through attachment–detachment of primary units to embryos of various sizes. The process can be represented by the following chain of reversible reactions:

$$\mathrm{X_1} + \mathrm{X_1} \underset{k_{2,1}}{\overset{k_{1,2}}{\rightleftharpoons}} \mathrm{X_2}$$

$$\mathrm{X_2} + \mathrm{X_1} \underset{k_{3,2}}{\overset{k_{2,3}}{\rightleftharpoons}} \mathrm{X_3}$$

$$\vdots$$

$$\mathrm{X_n} + \mathrm{X_1} \underset{k_{n+1,n}}{\overset{k_{n,n+1}}{\rightleftharpoons}} \mathrm{X_{n+1}}$$

$$\vdots$$

$$\mathrm{X_m} + \mathrm{X_1} \underset{k_{m+1,m}}{\overset{k_{m,m+1}}{\rightleftharpoons}} \mathrm{X_{m+1}} \quad m >> n^*.$$

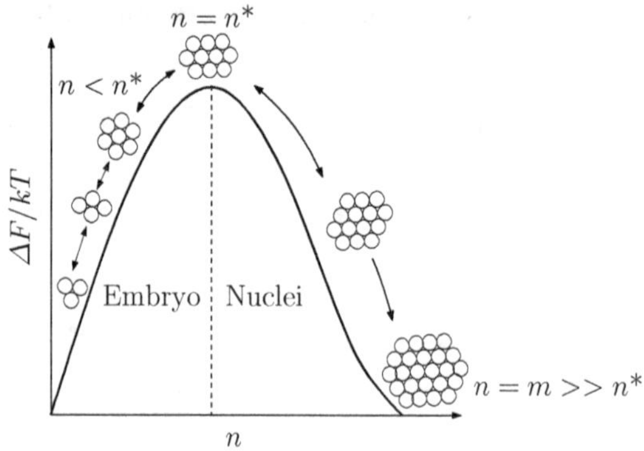

Figure 1.9 Embryos crossing the activation barrier to become nuclei.

In the above chain of reactions particles of size much larger than n^*, denoted by m, are also formed. As evident from Figure 1.8, the free energy change for these particles is negative and hence they grow at a very fast rate compared to particles around n^*. Hence, we can set m to a value such that dissolution of these particles (reverse reaction) is negligible and therefore the rate of formation of particles with m units can be considered as the rate of nucleation. This situation is shown pictorially in Figure 1.9.

The above chain of reactions is a larger version of series reactions encountered in chemical kinetics. In this case, the nuclei ($n >> n^*$) may be thought of as the product and other smaller species as intermediates. Like other series reactions, the intermediate concentration remains very small for the entire process in this case. Formation of an appreciable amount of product with a very small quantity of intermediate tells us that the chain reaches a steady state quickly.

In this chain of reactions, C_1 (concentration of X_1) depletes as the reaction proceeds, but the change is very small because the embryos are very small and cannot use up much of the material present. Hence, variation in C_1 may be neglected. But because there is a small variation in the concentration of C_1, we call this steady state the quasi steady state. We shall use this quasi steady state or pseudo steady state to evaluate the concentration of intermediates.

If we consider the species C_2, its net rate of production by the first reaction is

$$J_{1,net} = J_{1,2} - J_{2,1} = k_{1,2}C_1^2 - k_{2,1}C_2$$

and its net rate of consumption by the second reaction is

$$J_{2,net} = J_{2,3} - J_{3,2} = k_{2,3}C_1C_2 - k_{3,2}C_3.$$

If the concentration of C_2 remains constant, these two reactive fluxes must be equal. The same argument can be used for any intermediate species and this condition can be written as

$$J_{1,net} = J_{2,net} = J_{3,net} = \dots = J_{n,net} = \dots$$

Hence, we can drop the subscript and the net right ward flux can be generalized as

$$J_{net} = J_{n-1,n} - J_{n,n-1}$$

for any n. This applies even when $n >> n^*$. Since the net rate of production of particles of that size is same as the rate of nucleation, our task is to obtain J_{net}. Note that we want to express this flux in terms of the concentration of primary units (C_1) because this is the only measurable concentration. Hence, all other quantities should be eliminated. Let us start by expressing C_2 in terms of other variables:

$$J_{net} = J_{1,2} - J_{2,1} = k_{1,2}C_1^2 - k_{2,1}C_2.$$

Rearranging,

$$C_2 = \left(\frac{k_{1,2}C_1^2}{k_{2,1}} - \frac{J_{net}}{k_{2,1}}\right).$$

The above equation relates C_2 to C_1, but the rate constant remains unknown. Those can be obtained by using the equilibrium concentrations. Although the concentrations of the intermediate species are not accessible, their equilibrium concentrations (which are never reached during the process) are readily obtained in terms of measurable quantities (see eqn 1.12). *If the system attains equilibrium*, the rate constant remains the same, but the concentration reaches equilibrium values to exactly balance the forward and backward reaction fluxes. Denoting the equilibrium flux by J_{ij}^e,

$$J_{12}^e = J_{21}^e = k_{12}C_1^2 = k_{21}C_2^e.$$

Rearrangement of the above expression gives

$$\frac{k_{12}C_1^2}{k_{21}} = C_2^e$$

$$\frac{1}{k_{21}} = \frac{C_2^e}{J_{21}^e}.$$

Note that we did not use the superscript e for C_1 because its value remains practically constant during the tiny nucleation window. Using the above relations:

$$C_2 = C_2^e\left(1 - \frac{J_{net}}{J_{21}^e}\right).$$

We can also derive the expression for C_3 in a similar way:

$$C_3 = C_3^e\left(1 - \frac{J_{net}}{J_{21}^e} - \frac{J_{net}}{J_{32}^e}\right).$$

In a similar way, it can be shown that,

$$C_n = C_n^e \left(1 - \frac{J_{net}}{J_{21}^e} - \frac{J_{net}}{J_{32}^e} - \cdots - \frac{J_{net}}{J_{n,n-1}^e}\right).$$

If n is much larger than the critical nuclei size, as shown in Example 1.10, C_n^e becomes very high compared to other species. Hence, $C_n/C_n^e << 1$ when $n >> n^*$. Using this condition,

$$0 \approx \left(1 - \frac{J_{net}}{J_{21}^e} - \frac{J_{net}}{J_{32}^e} \cdots - \frac{J_{net}}{J_{n,n-1}^e}\right)$$

or the rate of homogeneous nucleation is given by

$$\dot{N} \approx J_{net} = \left(\frac{1}{J_{21}^e} + \frac{1}{J_{32}^e} \cdots + \frac{1}{J_{n,n-1}^e}\right)^{-1}. \tag{1.13}$$

Using the expression for flux and the equilibrium concentration from eqn 1.12,

$$J_{n,n-1}^e = k_{n,n-1} C_n^e = (k_{n,n-1}/v_m) \exp\left(\frac{n\phi - \psi n^{2/3}}{kT}\right).$$

In the above expression, everything except $k_{n,n-1}$ is known. Now, what could be a possible value of $k_{n,n-1}$? A useful approximation is to consider all the rate constants to be equal. We note that the rate constant is a first-order rate constant with unit of inverse time. A reasonable estimate of this time will be the characteristic time for the movement of molecules to the surface of the particle. If this process is limited by diffusion, the characteristic time will be $(D/d^2)^{-1}$, where d is the diameter of the molecules. Hence, although there could be exceptions, a reasonable estimate of $k_{n,n-1}$ is (D/d^2).

1.2.2.3 Algebraic Manipulation of the Rate Expression

Although we finished the conceptual development of the rate of homogeneous nucleation in the previous section, the final form is obtained only after extensive algebraic manipulation, which we will complete in this section. First, let us express the sum in eqn 1.13 in terms of measurable variables:

$$S = \frac{1}{J_{21}^e} + \frac{1}{J_{32}^e} \cdots + \frac{1}{J_{n,n-1}^e} = \sum_{n=2}^{m} k' \exp\left(\frac{-n\phi + \psi n^{2/3}}{kT}\right),$$

where

$$k' = \frac{v_m}{k_{n,n-1}} \approx \frac{v_m d^2}{D} \approx \frac{d^5}{D}.$$

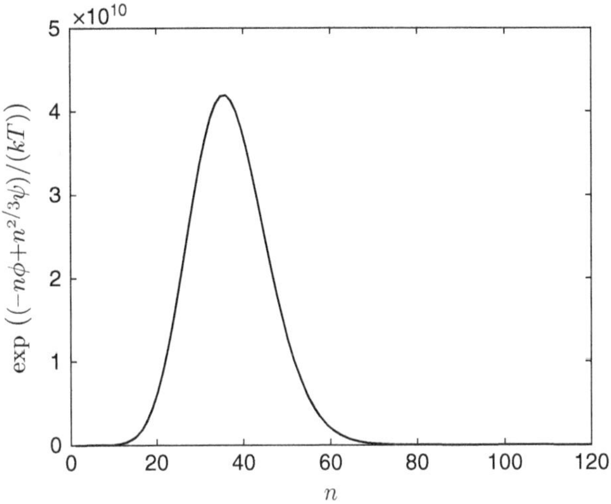

Figure 1.10 Significant portion of the integrand in eqn 1.14.

The increment in n may be considered small compared to the size domain which spans from $n = 2$ to $n = m >> n^*$. In that case, the sum can be approximated by an integral. We can also extend the lower limit to zero without affecting the accuracy significantly:

$$S = \int_{n=0}^{m} k' \exp\left(\frac{-n\phi + \psi n^{2/3}}{kT}\right) dn. \tag{1.14}$$

Even after this simplification, it is not very easy to integrate this exponential function. Hence, additional simplifications are required which will be clear after we examine the function closely. The function (integrand) is plotted for typical values of the parameters used in previous examples and is shown in Figure 1.10. For this set of parameters, the critical nuclei size is $n^* = 35$. It can be readily seen that the significant portion of the function lies between 0 and $n = 2n^*$. Hence, the limits of the integration may be changed to $[0 \quad 2n^*]$ without affecting the accuracy.

Now, using the relation $n^* = (2\psi/3\phi)^3$ to eliminate ψ, the sum becomes

$$\begin{aligned} S &= \int_{0}^{2n^*} k' \exp\left(\frac{-n\phi + \frac{3n^{*1/3}\phi}{2} n^{2/3}}{kT}\right) dn \\ &= \int_{0}^{2n^*} k' \exp\left(\frac{-\frac{n}{n^*}\phi n^* + \frac{3n^*\phi}{2}(\frac{n}{n^*})^{2/3}}{kT}\right) dn \\ &= \int_{0}^{2n^*} k' \exp\left[\frac{n^*\phi}{kT}\left(-\frac{n}{n^*} + \frac{3}{2}\left(\frac{n}{n^*}\right)^{2/3}\right)\right] dn. \end{aligned}$$

Denoting $n/n^* = 1 + x$,

$$S = k'n^* \int_{-1}^{1} \exp\left[\frac{n^*\phi}{kT}\left(-1 - x + \frac{3}{2}(1+x)^{2/3}\right)\right] dx.$$

Now, on expanding the power term and neglecting the higher order terms, the portion in the bracket becomes

$$-1 - x + \frac{3}{2}\left(1 + \frac{2}{3}x + \frac{-\frac{2}{3}\frac{1}{3}}{2}x^2 + \cdots\right) \approx \frac{1}{2} - \frac{x^2}{6}.$$

Therefore, the summation becomes

$$S = k'n^* \exp\left(\frac{n^*\phi}{2kT}\right) \int_{-1}^{1} \exp\left[-\frac{n^*\phi}{6kT}x^2\right] dx.$$

Now, since the integrand is practically zero beyond the limit $[-1 \quad 1]$, we can extend the limit of the integration to $-\infty < x < \infty$ without affecting its accuracy. We can readily see that this will cast the integral into the Gaussian form. Denoting $[(n^*\phi)/(6kT)]x^2 = y^2$,

$$\begin{aligned}
S &= k'n^* \exp\left(\frac{n^*\phi}{2kT}\right) \int_{-\infty}^{\infty} \exp(-y^2)\sqrt{\frac{6kT}{n^*\phi}}dy \\
&= k'n^* \sqrt{\frac{6kT}{n^*\phi}} \exp\left(\frac{n^*\phi}{2kT}\right) \int_{-\infty}^{\infty} \exp(-y^2)dy \\
&= k' \sqrt{\frac{6kTn^*\pi}{\phi}} \exp\left(\frac{n^*\phi}{2kT}\right).
\end{aligned}$$

Hence, the nucleation rate becomes

$$\dot{N} = S^{-1} = \frac{1}{k'}\sqrt{\frac{\phi}{6\pi kTn^*}} \exp\left(-\frac{\phi n^*/2}{kT}\right).$$

1.2.2.4 Various Forms of Homogeneous Nucleation Rate

The above expression for the homogeneous nucleation rate may be expressed in various forms. Because $\phi n^*/2 = \Delta F^*$, the rate can be written as

$$\dot{N} = \frac{D}{d^5}\sqrt{\frac{\phi}{6\pi kTn^*}} \exp\left(\frac{-\Delta F^*}{kT}\right). \tag{1.15}$$

In this case, the rate is expressed in terms of activation barrier, which is similar to the Arrhenius equation in chemical kinetics. It can also be written in terms of ϕ and ψ:

$$\dot{N} = \frac{3}{4}\frac{D}{d^5}\phi^2\sqrt{\frac{1}{\pi kT\psi^3}} \exp\left(-\frac{2\psi^3}{27kT\phi^2}\right). \tag{1.16}$$

In terms of supersaturation,

$$\dot{N} = \frac{3}{4}\frac{D}{d^5}[kT\ln(p'/p_0)]^2\sqrt{\frac{1}{\pi kT\psi^3}}\exp\left(-\frac{2\psi^3}{27(kT)^3[\ln(p'/p_0)]^2}\right). \tag{1.17}$$

Denoting $(\psi/kT)^3 = k_n$,

$$\dot{N} = \frac{D}{d^5}[\ln(p'/p_0)]^2\sqrt{\frac{9}{16\pi k_n}}\exp\left(-\frac{(2/27)k_n}{[\ln(p'/p_0)]^2}\right).$$

This is often written as

$$\dot{N} = k_{n1}[\ln(p'/p_0)]^2\exp\left(-\frac{k_{n2}}{[\ln(p'/p_0)]^2}\right). \tag{1.18}$$

The exponential term has the strongest influence on the nucleation rate. For this reason, the pre-exponential factor is often taken to be a constant. This leads to a more familiar form of homogeneous nucleation rate:

$$\dot{N} = k_{n1}\exp\left(-\frac{k_{n2}}{[\ln(p'/p_0)]^2}\right).$$

Example 1.11 For water vapour, compute the values of k_{n1} and k_{n2}. Calculate the nucleation rate for supersaturation ratios of 4 and 1.8.

Solution: The value ψ/kT has been computed earlier to be 6.8. Hence, $k_n = 314$, which gives k_{n2} as 23.3 and

$$k_{n1} = \frac{1\times10^{-5}}{10^{-50}}\sqrt{\frac{9}{16\pi\times314}} \approx 2.4\times10^{43}.$$

For a supersaturation ratio of 4, $[\ln(p'/p_0)]^2 \approx 2$. The nucleation rate becomes

$$\dot{N} \approx 2\times2.4\times10^{43}\exp\left(-\frac{23.3}{2}\right) \approx 4.2\times10^{38} \quad \text{nuclei/m}^3\text{/s}.$$

For a supersaturation ratio of 1.8 this becomes

$$\dot{N} \approx 0.35\times2.4\times10^{43}\exp\left(-\frac{23.3}{0.35}\right) \approx 1\times10^{14} \quad \text{nuclei/m}^3\text{/s}.$$

It is clear from the above example that the rate of homogeneous nucleation is a very strong function of supersaturation. It changes by 24 orders of magnitude when the supersaturation ratio changes merely by a factor 2.2! It can also be noted that the pre-exponential factor only changes a little compared to the exponential term on changing the supersaturation. Hence, for most practical purposes, the pre-exponential term may be considered independent of supersaturation. □

1.2.3 Experimental Aspects of Homogeneous Nucleation

The homogeneous nucleation theory or classical nucleation theory (CNT) is very popular among the scientific and technical community. The reason for this is clear from the above examples: CNT predicts a very rapid increase in the nucleation rate beyond a critical supersaturation. This is observed in all systems which nucleate homogeneously [4]. Such a catastrophic break out of nucleation is not predicted by a power law or other simpler expressions. For water vapour, this calculated critical supersaturation is $p'/p_0 = 4.2$, which matches observations very well.

To achieve homogeneous nucleation experimentally, the critical step is to remove all possible particles from the system which may lead to heterogeneous nucleation. We shall see shortly that heterogeneous nucleation has a lower free energy barrier and hence happens long before homogeneous nucleation. Hence, if dust particles are present, homogeneous nucleation cannot be observed. Another challenge for such experiments is to detect the nuclei. Fortunately, small particles of many materials are good scatterers of visible light and very small particles can be detected using light-scattering techniques.

1.2.3.1 Investigation Using a Cloud Chamber

One of the tools that is used for experimental investigation of homogeneous nucleation is the diffusion cloud chamber. A schematic of this machine is shown in Figure 1.11. The key function of the diffusion cloud chamber is to maintain a known supersaturation at a given location. The occurrence of homogeneous nucleation is detected using laser light scattering and hence the critical supersaturation at which nucleation starts can be determined readily.

The equipment consists of a cylindrical chamber with a top plate and a bottom plate whose temperature can be controlled very accurately. The bottom plate is maintained at a higher temperature than the top plate and a pool of liquid is maintained on the bottom plate. The chamber also contains an inert gas. Because the top plate is colder, a portion of water vapour condenses on it. This sets a concentration gradient inside the chamber and vapour molecules diffuse towards the top plate through the non-diffusing gas molecules. The top plate is bevelled, as shown in the Figure 1.11, so that the condensed drops drip through the wall instead of falling through the gas phase like rain. The walls are made of insulating quartz glass so that the chamber can be seen without any heat loss.

The steady state temperature and concentration profiles can be used to obtain the supersaturation profile, as shown in Figure 1.11. The maximum supersaturation occurs somewhere near the middle (its exact location can be calculated). The magnitude of this supersaturation can be altered by changing the temperatures of the top and bottom plates.

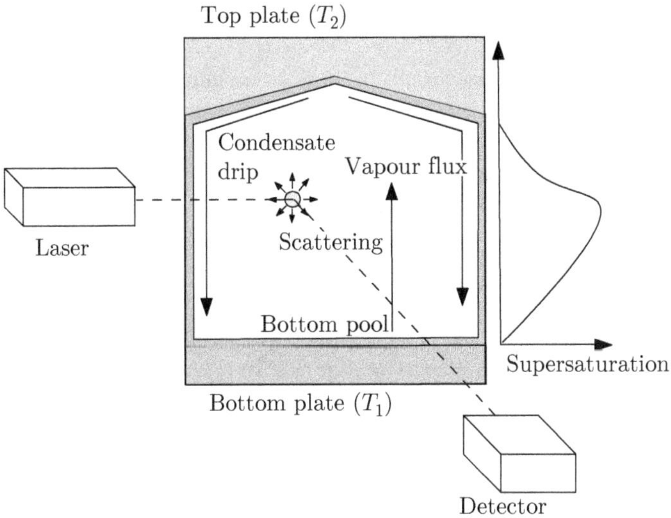

Figure 1.11 Experimental determination of the critical supersaturation of homogeneous nucleation using a diffusion cloud chamber [6, 7].

At the beginning, the supersaturation that can be reached is small due to dust particles present in the system. These particles induce heterogeneous nucleation and thereby relieve the supersaturation. However, each time a dust particle participates in the nucleation, it precipitates with the liquid drops and is removed from the vapour space. This cleans up the apparatus of these disturbances.

When the system is cleaned, the supersaturation can reach a very high value without nucleation. It also shows the feature that slight increase in supersaturation from a certain value leads to nucleation in a burst. The vapour phase suddenly becomes cloudy when the maximum supersaturation is reached. Although the naked eye may not be able to detect the cloudiness, the generation of a large number of small particles contributes to light scattering, which is detected by a sensitive detector located at 90°. Similar phenomena also occurs during formation of cloud in nature. For details see Ref. [5].

1.2.3.2 Other Methods

A number of other methods are also available to monitor the rate of nucleation. In one such method, the system is allowed to nucleate for a short period at a prescribed supersaturation (higher than the critical one). Then the supersaturation is quickly brought below the critical value required for nucleation but kept higher than saturation. Under this condition no new nuclei can form but the existing ones can grow. When the particles grow to an observable size, they may be counted using a microscope.

1.3 Non-Homogeneous Nucleation

It has been shown in the previous section that homogeneous nucleation requires a very high supersaturation. It has been observed that in many cases new particles nucleate at a much lower supersaturation than that required for homogeneous nucleation. This happens because other routes for nucleation exist. One of them is heterogeneous nucleation.

1.3.1 Heterogeneous Nucleation

Nucleation is much easier on a foreign surface. The surface reduces the energy barrier needed for nucleation. To calculate the net free energy change for heterogeneous nucleation, let us consider a flat surface on which a small cap of liquid nucleates, as shown in Figure 1.12. It is clear from Figure 1.12 that in contrast to homogeneous nucleation, the heterogeneous nuclei contain both solid–liquid and liquid–vapour interfaces and the solid–liquid interface is formed by eliminating a portion of the solid–vapour interface. Hence, the free energy change for this case can be written as

$$\Delta F = -\frac{V_C}{v''} RT \ln\left(\frac{p'}{p_0}\right) + A_c\gamma_{lv} + A_b\gamma_{sl} - A_b\gamma_{sv}.$$

The volume of the cap V_C, the base area A_b, and the cap area A_c are functions of the cap radius (r) and contact angle θ. The contact angle is a system-specific property and is constant for a given solid–liquid vapour system. If the contact angle is very small, the liquid tends to form a film on the solid and is called a wetting liquid with respect to the solid. If the contact angle is very large, the drop becomes almost spherical and the liquid is non-wetting.

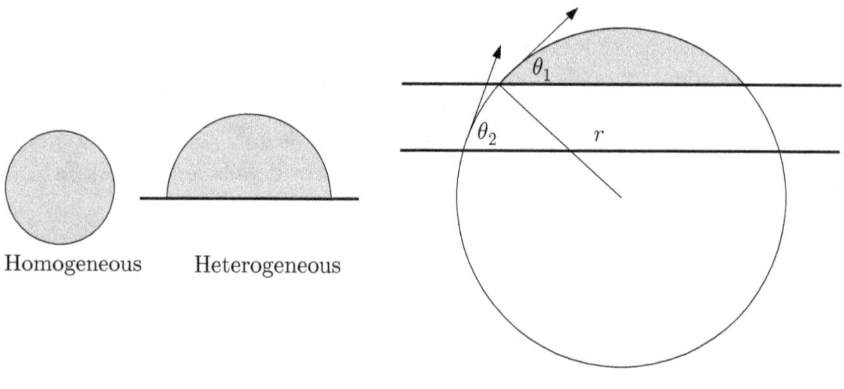

Figure 1.12 Left: homogeneous vs heterogeneous drops. Right: drop volume as a function of cap radius and contact angle. The same cap radius may produce a different drop volume if the contact angle is different.

The three quantities, V_C, A_b, and A_c, are readily expressed in terms of r and θ, and are given by (see section A.1 for derivation)

$$V_C = \frac{\pi r^3}{3}(1 - \cos\theta)^2(2 + \cos\theta)$$

and

$$A_b = \pi r^2(1 - \cos^2\theta), \quad A_c = 2\pi r^2(1 - \cos\theta).$$

Now, the three surface energies, γ_{lv}, γ_{sv}, and γ_{ls}, are not independent. They are related by a force balance (see section A.2 for derivation) given by

$$\gamma_{lv}\cos\theta = \gamma_{sv} - \gamma_{ls}.$$

Substituting these relations and after some algebraic manipulation, the free energy of formation of the cap becomes

$$\Delta F = -\frac{V_C}{v''}RT\ln\left(\frac{p'}{p_0}\right) + \gamma_{lv}\pi r^2(1 - \cos\theta)^2(2 + \cos\theta).$$

The critical free energy and cap radius can be readily calculated as

$$\Delta F^* = \frac{4\pi(r^*)^2}{3}\gamma_{lv}\frac{(1 - \cos\theta)^2(2 + \cos\theta)}{4} = \frac{4\pi(r^*)^2}{3}\gamma_{lv}f(\theta) \tag{1.19}$$

and

$$r^* = \frac{\gamma_{lv}v''}{RT\ln(p'/p_0)}.$$

It can be seen that the critical free energy barrier is very similar to that for homogeneous nucleation except for the part involving the contact angle. It can also be seen that the contact angle is a key factor in determining the energy barrier to the process. For example, if the contact angle is zero, the energy barrier simply vanishes! This means that even for a very small supersaturation, a liquid phase will form and will be deposited on the surface as a *film*. On the other hand, if the liquid is completely non-wetting, θ will be 180° and the energy barrier will be same as that for homogeneous nucleation. For intermediate cases, the energy barrier will be between zero and that of homogeneous nucleation. To find the rate of heterogeneous nucleation, one would have to substitute ΔF^* and the corresponding n^* from eqn 1.19 into eqn 1.15.

Example 1.12 What is the rate of nucleation of water vapour if graphite particles are present for a supersaturation ratio of 1.8? The contact angle is 86°.

Solution: In this case, $f(\theta) = 0.45$. Therefore, the nucleation rate becomes

$$\dot{N}_{Het} \approx 1 \times 10^{43}\exp\left(-\frac{23.3 \times 0.45}{0.35}\right) \approx 1 \times 10^{30}.$$

It can be seen that a 16 orders of magnitude increase in nucleation rate is possible if graphite particles are present. □

1.3.2 Nucleating Agents and Organizers

Several substances are found to assist in nucleation. Such substances are called nucleating agents. Polymer nucleation is often assisted by such substances. For example, at practically attainable supersaturation, the nucleation rate of polypropylene is very low but the growth rate is very high. Hence, the supersaturation is quickly exhausted soon after the formation of a small number of nuclei and large crystallites are formed. To produce particles of a desired size, the nucleation rate must be boosted. This is carried out by using a nucleating agent called dibenzylidine sorbitol [8]. Similarly, nucleation of ice can be promoted by silver iodide. Silver iodide has a crystal structure very similar to ice. Hence, water molecules consider the silver iodide particles to be ice nuclei and start growing on all such centres.

Another type of nucleating agent are called "organizers" [9]. Several ions and large molecules can act as organizers for nucleation. Such molecules bring many atoms together and create a very high local supersaturation even though the average supersaturation may be low. As a result, nucleation can start at much lower supersaturation. Tannin has been shown to act as an organizer for the synthesis of gold nanoparticles [9].

1.3.3 Secondary Nucleation

The three mechanisms discussed above are called primary nucleation because they produce particles from a particle-free solution. However, there are other routes of nucleation that are particularly important in industrial contexts where new particles are formed in the presence of other particles [10]. These routes are called secondary nucleation. Secondary nucleation occurs at a much lower supersaturation. There are a few well-known theories that explain the formation of new particles from existing particles:

1) Initial breeding or dust breeding: Often particles are added to the crystallizer as seed. Such seed particles usually contain small particles on their surfaces. In solution, these particles dislodge from the surface and start growing. The rate of nucleation for this case must be independent of supersaturation and stirrer speed.
2) Needle breeding: At high supersaturation, many particles grow into needle-shaped crystals. Such needles are readily fractured by hydrodynamic force and give rise to smaller particles that contribute to secondary nucleation.
3) Polycrystalline breeding: Many fast-growing crystals show large numbers of defects and amorphous zones. These imperfect crystals may break into smaller fragments to produce particles.
4) Micro-abrasion: At high stirring speed, the sharp corners of the crystals break by colliding with each other and give rise to small crystallites. If

micro-abrasion is present, the crystals look like rounded particles and the nucleation rate is usually dependent on the hardness of the material.

5) Dendritic growth and fluid shear: Some crystals tend to grow dendritic structures which break under fluid shear in a similar way as in needle breeding.
6) Contact nucleation: The contact of a particle with vessel wall, stirrer or other particles leads to contact nucleation in a similar way to micro abrasion.

A generalized theory like homogeneous nucleation does not exist for secondary nucleation. Empirical relationships such as

$$\dot{N}_{Sec} = k_N M_T^j \Delta C^n$$

are usually used. Here M_T is the magma density, i.e the mass of particles per unit volume of the slurry, and ΔC is the supersaturation. n and j are empirically fitted constants.

1.4 Exercises

Exercises 1.1 Consider 1 l of a *saturated* salt solution at room temperature. To this solution, 50 g of 0.5 μm particles of the salt is added. After a few minutes, 25 g of 0.1 μm particles is also added. Will there be any change in the system with time? Explain your answer *quantitatively*. Data: solubility of the salt at room temperature 200 g/l, surface energy of the salt–water interface 500 mJ/m^2, density and molecular weight of the salt 3 g/cc and 150 g/mol, respectively. Will the observation be different if only 8 g of smaller particles is added?

Exercises 1.2 Consider the system shown in Figure 1.13. The nozzle is a very narrow capillary tube. If the gas is blown at a very small velocity (so that the bubble forms and grows slowly), the pressure difference measured by the manometer is observed to go through a maximum. Can you explain this observation?

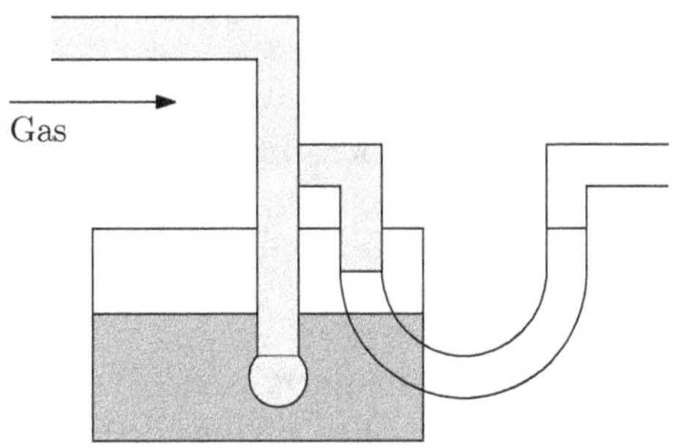

Figure 1.13 Slow formation of a bubble on the tip (Exercise 1.2).

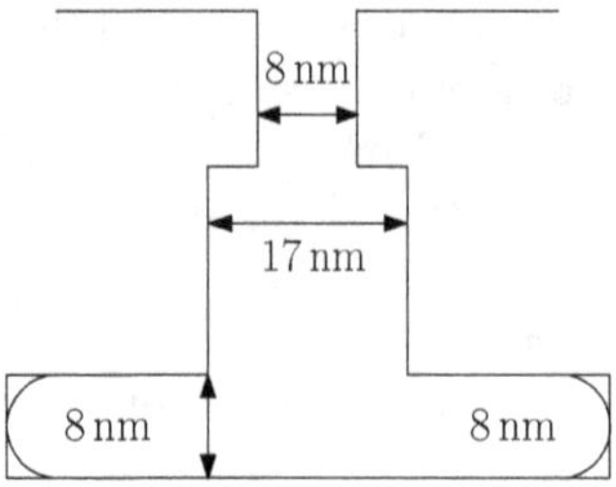

Figure 1.14 A complex cavity (Exercise 1.6).

Exercises 1.3 A subnanometre condensed water is formed due to thermal fluctuation in a wedge-shaped nano-crack present on a hydrophilic surface. The prevailing supersaturation ratio is 1.0. Will the liquid phase grow?

Exercises 1.4 In heterogeneous nucleation of water on soot particles, the nucleation rate may be dependent on the number density of the particles present. Analyse this statement using the nucleation model.

Exercises 1.5 Explain the organizer mechanism of nucleation in detail. Search the current literature for more information. Can you obtain a rate expression for organizer-based nucleation?

Exercises 1.6 A hydrophilic surface is exposed to pure dust-free water vapour at 100 °C. Water condenses at the corners of the cavity, as shown in Figure 1.14, due to thermal fluctuation when the prevailing supersaturation ratio is 0.9. The supersaturation ratio is then increased to 0.95 and after sufficient time reduced to 0.9. Discuss the evolution of the system.

Exercises 1.7 In many systems, negligible nucleation is observed even when very high supersaturation is created. This phenomenon is related to the degree of cooling versus the rate of nucleation where the supersaturation is created by cooling the system. The nucleation rate reaches a maximum at a certain supersaturation and then drops quickly. Because of cooling, p_0 is changed according to Clapeyron's equation:

$$\ln \frac{p}{p_0} = \frac{\Delta H_v}{R}\left(\frac{1}{T} - \frac{1}{T_0}\right).$$

Using the above expression show that the nucleation rate reaches a maximum value at a certain supersaturation. Express this supersaturation as a function of measurable quantities and then explain the phenomenon observed using this theory.

Exercises 1.8 Examine the current literature and list all possible modes of secondary nucleation. Explain the physics of each process with diagram.

Bibliography

1 A. W. Adamson and A. P. Gast. *Physical Chemistry of Surfaces*, 6th edition. John Wiley & Sons, 1997.
2 R. J. Hunter. *Foundations of Colloid Science*, 2nd edition. Oxford University Press, 2001.
3 A. E. Nielsen. *Kinetics of Precipitation*. Pergamon Press, Oxford, 1964.
4 V. K. La Mer. Nucleation in phase transitions. *Industrial & Engineering Chemistry*, *44*(6):1270–1277, 1952.
5 S. K. Friedlander. *Smoke, Dust and Haze: Fundamentals of Aerosol Behavior*. John Wiley & Sons, 1977.
6 D. C. Marvin and H. Reiss. Photooxidation of sulfur dioxide. *Journal of Chemical Physics*, *69*(5):1897–1918, 1978.
7 J. Smolík and V. Ždímal. Condensation of supersaturated vapors of dioctylphthalate: Homogeneous nucleation rate measurements. *Aerosol Science and Technology*, *20*(1):127–134, 1994.
8 Y. Feng, X. Jin, and J. N. Hay. Effect of nucleating agent addition on crystallization of isotactic polypropylene. *Journal of Applied Polymer Science*, *69*(10):2089–2095, 1998.
9 S. Kumar, K. S. Gandhi, and R. Kumar. Modeling of formation of gold nanoparticles by citrate method. *Industrial & Engineering Chemistry Research*, *46*(10):3128–3136, 2007.
10 A. S. Myerson. *Handbook of Industrial Crystallization*, 2nd edition. Butterworth-Heinemann, 2002.

2

Growth

We saw in the previous chapter that nucleation occurs in a burst as soon as the supersaturation reaches a critical value. Nucleation continues until the supersaturation is brought below the critical limit by exhausting the material. Can nucleation bring down the supersaturation and be self limiting? Typical numbers show that nucleation alone cannot consume enough material to change the supersaturation significantly. Particle growth is responsible for bringing down the supersaturation and limiting the period of nucleation to a tiny time window.

Let us consider the case of water vapour nucleating at a supersaturation ratio of 2.5. At this supersaturation ratio the nucleation rate will be 8×10^{19} nuclei/m^3/s whereas approximately 4×10^{25} vapour molecules will be present in that volume. Hence, only a tiny fraction of material (2×10^{-6}) is consumed per second. So, to bring the concentration down to 90% of its initial value will take about 10^6 s. It is clear that if nucleation is the only event bringing down supersaturation, the system needs to keep nucleating for days. Such prolonged nucleation is never observed and, in general, nucleation finishes very quickly, within a fraction of a second. The phenomenon responsible for bringing down the supersaturation is growth. The nuclei offer a surface on which significant growth occurs to bring down the supersaturation below the critical value needed for nucleation.

Because nucleation happens at high supersaturation, very fast growth of the nuclei occurs at the nucleating supersaturations. More growth adds more area to the existing particles and a combination of nucleation and growth brings down the supersaturation below the critical value needed for nucleation very quickly, usually within a fraction of a second. Nucleation practically ceases after this short burst but growth continues for a much longer period, typically for minutes to hours. The change in supersaturation during nucleation and growth is shown schematically in Figure 2.1. In this chapter we shall concentrate on the rate of growth of particles.

Engineering of Submicron Particles: Fundamental Concepts and Models, First Edition. Jayanta Chakraborty.

Companion website: http://booksupport.wiley.com

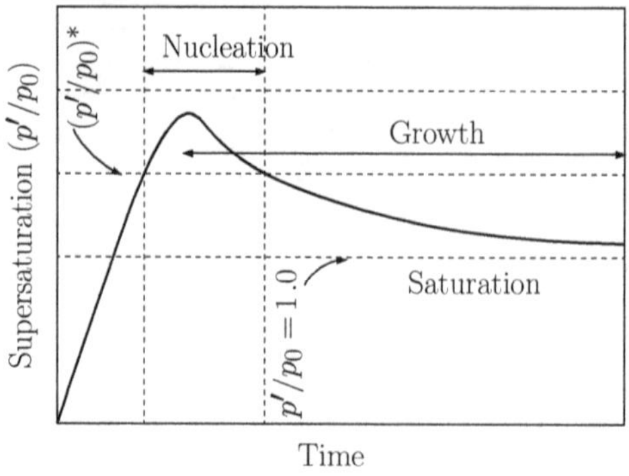

Figure 2.1 Build up and exhaustion of supersaturation during the nucleation growth process. The supersaturation below which nucleation proceeds at a vanishingly small rate is denoted by $(p'/p_0)^*$.

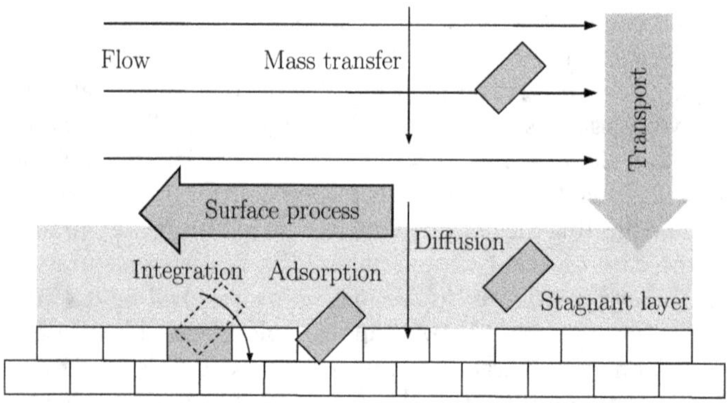

Figure 2.2 Transport, adsorption, and integration of the growth unit to the growing crystal surface.

2.1 Traditional Crystal Growth Models

It is advantageous to visualize the growth process as lying bricks on a plane surface where the bricks are brought to the surface through a multi-step process, as shown in Figure 2.2 [1]. Growth occurs on a solid surface in a circulating fluid medium and the growth units must travel towards the surface by convection and diffusion. The presence of a solid surface also induces a stagnant layer close to the surface. Hence, the growth species must be transferred to the surface by convection in the flow region and through diffusive movement near the surface. Next, the growth unit must be adsorbed on the surface and finally integrated into the crystal lattice. The last two steps are often called surface integration.

Depending on the system and growth conditions, either the transport process or the surface integration becomes rate limiting. The mass transfer in the flowing fluid is usually not the limiting process and transport-limited growths are mostly diffusion controlled. Surface integration may also become rate limiting.

The surface integration controlled growth may be classified into surface nucleation controlled growth and surface dislocation controlled growth depending on the mechanism of surface integration. We shall discuss these three growth mechanisms in this chapter. In all three cases, we shall assume the crystal to be spherical or quasi spherical and will predict the linear growth velocity dR/dt as a function of particle size and supersaturation.

2.1.1 Diffusion Controlled Growth

Let us consider a spherical particle of radius R growing inside a fluid with a spherically symmetric supersaturation field, as shown in Figure 2.3. The particle is immersed in the fluid of uniform concentration c_∞ at $t = 0$ and a transient concentration profile is established as a result of growth. For spherical geometry, the transient mass balance equation for purely diffusive transport may be written as [2]

$$\frac{\partial c}{\partial t} = D\frac{1}{r^2}\frac{\partial}{\partial r}\left(r^2\frac{\partial c}{\partial r}\right).$$

Usually the development of the concentration profile is very quick in compared to the growth of the particle. This fact can be used to simplify the model substantially by assuming quasi steady state of the concentration profile:

$$\frac{d}{dr}\left(r^2\frac{dc}{dr}\right) = 0.$$

Since diffusion is the rate-controlling step, surface integration is very fast. Hence, the material near the surface is consumed quickly until the saturation concentration is reached. Hence, the boundary conditions can be written as

$$@r = R, c = c_{sat}; \qquad @r = \infty, c = c_\infty$$

and the solution becomes

$$c = c_\infty - \frac{(c_\infty - c_{sat})R}{r}.$$

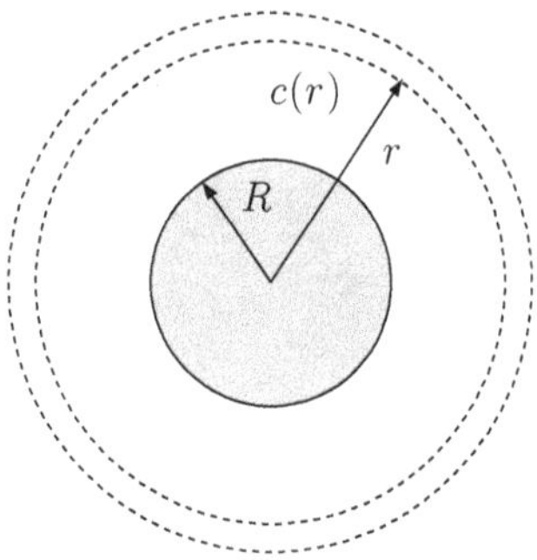

Figure 2.3 Mass balance in a spherical shell around the growing spherical particle.

Note that we can let R vary with time after we obtain the concentration profile using the quasi steady state assumption. Now, equating the rate of mass transport to the rate of mass gain by the particle:

$$\frac{d}{dt}\left[\frac{4}{3}\pi R^3 \rho_p\right] = 4\pi R^2 D \left.\frac{dc}{dr}\right|_{r=R}.$$

Simplifying:

$$\frac{dR}{dt} = \frac{D(c_\infty - c_{sat})}{\rho_p R}. \tag{2.1}$$

It can be seen that for diffusion controlled growth the growth rate is inversely proportional to the particle size. Hence, smaller particles grow (or dissolve) at a much faster rate than larger particles. Thus, if a polydisperse set of particles are grown under diffusion controlled conditions, the size distribution will gradually narrow down. This phenomenon is called *size focusing* and is observed in many cases.

If surface integration becomes rate controlling instead of diffusion, the consumption of material near the surface becomes very slow compared to diffusive flux. This means that at steady state, when the two fluxes are equal, a small concentration gradient between the bulk and the surface will be able to support the consumption of material at the surface. This means that the surface concentration will not be depleted to c_{sat}. Rather it will be close to c_∞. This fact will be very useful in the next section where we will discuss surface nucleation controlled growth.

2.1.2 Surface Nucleation Controlled Growth

The growth of an existing particle occurs by condensation because the condensed phase is energetically favorable. We have seen in the case of particle nucleation that this negative free energy is countered by the positive surface energy term of a spherical nuclei. A similar phenomenon occurs when a small cluster tries to form on a plain surface, as shown in Figure 2.4.

Let us consider a flat surface and imagine the growth units as cubes. Each surface of the cube can form a bond and formation of bonds lowers the free energy

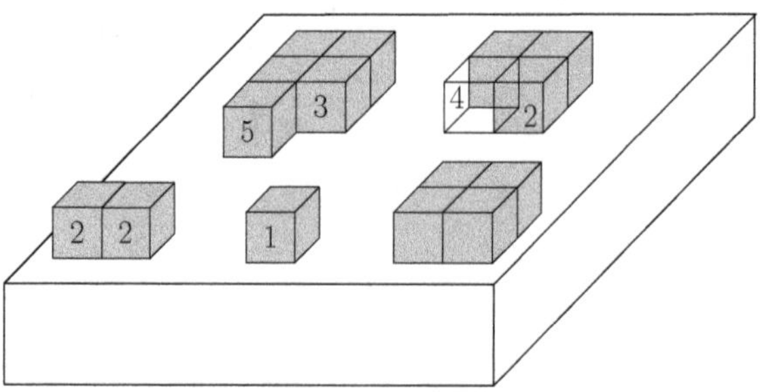

Figure 2.4 Formation of surface nuclei.

if the surrounding fluid is supersaturated. When an individual cube attaches to the surface, it can make only one bond, to the bottom surface, and very little volume free energy is gained. But it has to pay a penalty in terms of 'edge energy' for all four unattached sides. The overall free energy change is positive and hence such units leave the surface very quickly.

If a second cube attaches to the first one, the unit has a total of six exposed sides instead of eight and the gain in volume free energy is double. The situation is better than a single unit. Units attaching at position 3 or 4, as shown in Figure 2.4, will have even less edge energy for one unit of gain in free energy. It is clear that, like formation of nuclei in a three-dimensional space, the size of the surface cluster dictates the edge energy of the cluster: the smaller the surface cluster, the higher its edge energy. Hence, surface clusters smaller than a critical size will tend to evaporate and larger ones will grow to cover the surface. In line with the terminology used in the previous chapter, a surface cluster that may grow or dissolve is termed an embryo and a cluster that is large enough to grow is termed a surface nucleus.

If a surface nucleus is present, other units will attach to it in the 'steps'. Two such attachments are shown in Figure 2.4 as units 3 and 4. These units have only two unbounded faces and hence their formation is energetically favorable compared to that of 2 or 5. Usually the limiting process is the formation of surface nuclei and as soon as surface nuclei are formed on the surface they grow very quickly to cover the entire surface. In some cases a number of surface nuclei are formed and start to spread on the surface. Such a situation is shown in Figure 2.5. This is called polynuclear growth. In many cases a simpler situation is observed. If the rate of spread of the surface nucleus is much faster than

Figure 2.5 Micrographs of two-dimensional nucleation: polynuclear growth [3].

the surface nucleation rate, one layer is laid immediately after the first nucleus is formed. This is called mononuclear growth. For mononuclear growth, the surface advances by one molecular layer for each nucleation event.

2.1.2.1 Rate of Mononuclear Growth

Let us consider a circular surface nucleus containing n units on a plane surface. The free energy change of formation of this cluster is

$$\Delta F = -n\phi + L\sigma, \tag{2.2}$$

where ϕ is the free energy gain per unit by formation of the cluster, σ is the 'edge energy', and L is the circumference of the surface nucleus. Because we are interested in the rate of surface nucleation, we need to obtain the critical free energy for the process. Minimizing the change of free energy:

$$\frac{d\Delta F}{dn} = -\phi + \sigma\frac{dL}{dn}. \tag{2.3}$$

Now, if we define a as the area corresponding to one unit, the perimeter of the spot containing n units is given by $L = 2\sqrt{\pi a n}$. Therefore

$$\frac{dL}{dn} = \sqrt{\frac{\pi a}{n}}.$$

Substituting this expression in eqn 2.3 and setting $d\Delta F/dn = 0$ we obtain

$$n^* = \frac{\sigma^2 \pi a}{\phi^2} \tag{2.4}$$

and substituting n^* in eqn 2.2,

$$\Delta F^* = n^*\phi.$$

The rate of surface nucleation is therefore given by

$$\dot{N}_s = k_1 \exp(-\Delta F^*/kT) = k_1 \exp\left(\frac{-n^*\phi}{kT}\right).$$

For isothermal condensation of a vapour under pressure supersaturation, the affinity function ϕ can be written as $kT\ln(p/p_0)$. If the growth of particles in solution is of interest, it can be expressed as

$$\phi = kT\ln(c/c_s).$$

Note that for surface nucleation, the nucleation rate is given by number of nuclei born/unit area/unit time.

Now let us see how the growth rate of a crystal can be computed from the surface nucleation rate. Because growth is surface nucleation controlled, plenty of growth units are available near the surface and wait for a surface nucleation event. As soon as a surface nucleus (not an embryo) is born, the growth unit attaches to it very quickly and fills the surface with one complete layer. The newly formed surface then waits for the next nucleation to occur. Hence, the surface advances by one molecular dimension (d) for each nucleation event. It should be noted that the number of nucleation events occurring per unit time is dependent on the amount of area available.

Hence, the normal direction growth rate is given by

$$\frac{dR}{dt} = \dot{N}_s \times A \times d,$$

where A is the area of the crystal surface and d is the thickness of one atomic layer. As a general form, this can be written as

$$\frac{dR}{dt} = C \times \dot{N}_s(c/c_{sat}) \times R^2.$$

It can be seen that for surface nucleation controlled growth, larger particles have a higher linear growth rate than smaller particles. Hence, the size distribution broadens as growth progresses.

Example 2.1 Peng et al. [4] investigated the growth of semiconductor nanocrystals and observed the evolution of average size and polydispersity, as shown in Figure 2.6. Explain the observation using growth theories.

Solution: There is a clear size focusing at the early stage of growth. This is a clear indication that early growth is diffusion controlled. Because the particle size is small, the diffusive flux is very high and growth is very rapid, as seen in the figure. As particles grow in size, the rate of diffusion and hence the growth rate both diminish. After about 80 minutes the growth almost stops and the distribution broadens. This is because at this point the supersaturation is exhausted. Hence, particles that are slightly smaller dissolve and particles that are slightly larger grow. Thus, the size distribution broadens. □

2.1.3 Surface Dislocation Controlled Growth: BCF Theory

Often it is observed that crystals grow at a reasonable rate at a low supersaturation. If the surface nucleation rate at those supersaturations is calculated, it turns out to be essentially zero. While investigating the reason for such growths, Burton, Cabrera, and Frank found that growth at such low supersaturation is facilitated by the presence of screw dislocations on crystal surfaces. A screw dislocation is a kind of crystal defect, as shown in Figure 2.7.

As discussed for the case of surface nucleation, it is a lot easier for a growth unit to deposit beside a step than as an isolated unit. For this reason the surface nucleus grows at a very fast pace to cover the entire surface with a molecular layer. However, formation of a complete layer destroys the step, and growth must stop and wait for the next nucleation.

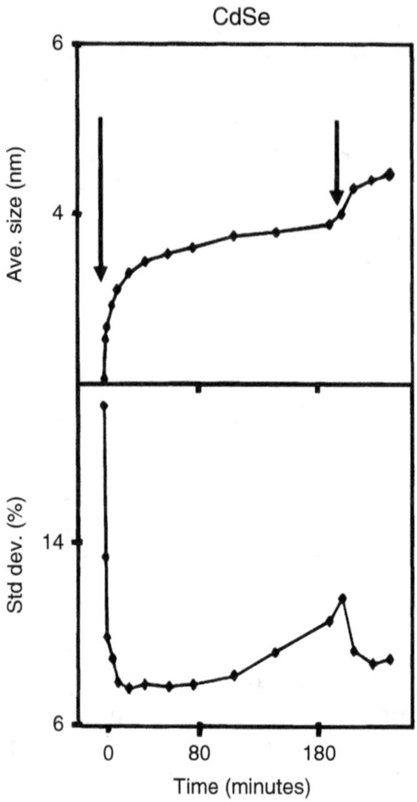

Figure 2.6 Growth of semiconductor nanocrystals (data for Example 2.1).

If a screw dislocation is present, it provides a step which does not vanish after growth of an atomic layer. This is clear from Figure 2.7. If we add one layer to the step (the grey unit), the step is retained. Even if we keep advancing towards the edge and hit the edge, another step normal to the previous one is created. At low supersaturation the edge growth rate is of course slower but the waiting time between two subsequent nucleation events is gone. Hence the growth proceeds at a reasonable rate even at low supersaturations. A growth spiral for a lysozyme crystal is shown in Figure 2.8.

2.1.3.1 Rate of Surface Dislocation Controlled Growth

The rate of surface dislocation controlled growth is determined by the rate of advancement of the spiral. First, let us concentrate on estimating the normal velocity of the spiral.

We shall conduct an approximate analysis to figure out the size dependance of this growth mechanism. We shall try to estimate the growth rate by matching the material consumed by growth with the diffusive flux of material towards

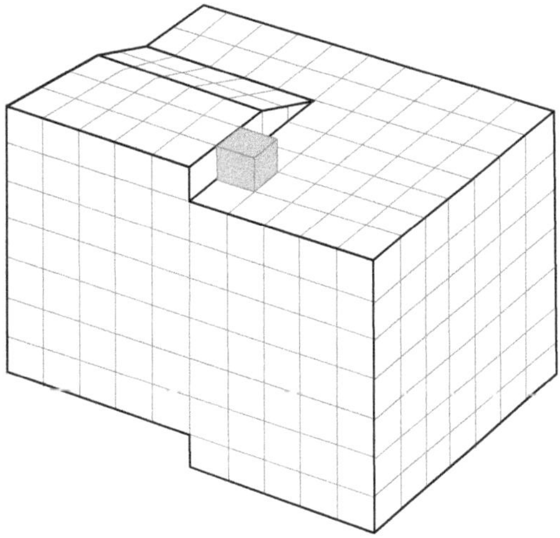

Figure 2.7 Screw dislocation.

Figure 2.8 Micrograph of growth spirals [3].

a spiral step (see Figure 2.9) where the radius of curvature is r. Note that irrespective of whether the process is diffusion limited or not, the diffusive flux and surface integration flux must match under steady state growth conditions. An order of magnitude estimate of the diffusive flux is

$$D(c_\infty - c_r)/L,$$

where c_∞ is the bulk concentration, c_r is the concentration near the spiral with local radius r, and L is the characteristic dimension of the crystal. It is clear from the figure that the area which receives the diffusive flux is $L \times d$ and hence the

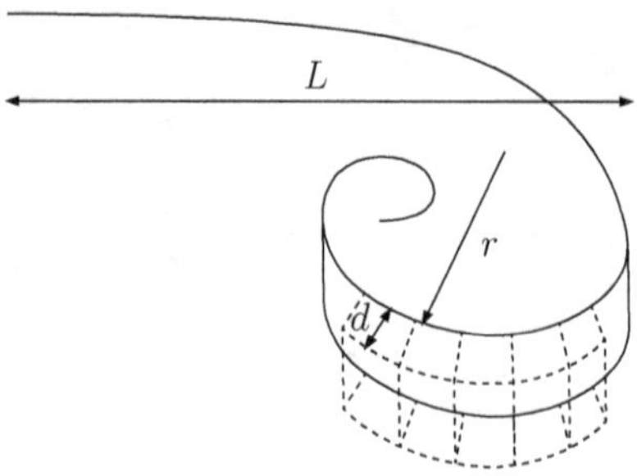

Figure 2.9 Schematic of spiral growth.

mass transfer rate towards the spiral step is $D(c_\infty - c_r)d$. Usually, the diffusivity is in molar units and the mass transfer rate in terms of number of units is given by $ND(c_\infty - c_r)d$, where N is the Avogadro number. Because the addition of each unit advances the spiral by length d, the velocity of the advancement of the spiral is

$$u = NDd^2(c_\infty - c_r).$$

Now we have to estimate the concentration c_r. c_r should be lower than c_∞ and higher than c_{sat}. It should also be able to grow a step which exists on a curved edge of radius r. Hence, it cannot be lower than the concentration that is in equilibrium with a critical surface nucleus of radius r. Using eqn 2.4 and recognizing that $an^* = \pi r^{*2}$, we obtain

$$r^* = \frac{\sigma a}{\phi}.$$

Substituting $\phi = kT\ln(c_r/c_{sat})$, we obtain

$$c_r = c_{sat}\exp(\sigma a/kTr),$$

which finally gives

$$u = NDd^2(c_\infty - c_{sat}\exp(\sigma a/kTr)).$$

Now we have the outward velocity of the spiral as a function of supersaturation and its curvature. It can be seen that this velocity is independent of the size of the crystal (L) and hence the dislocation controlled growth is size independent. Formally,

$$\frac{dR}{dt} = f(c).$$

The progress of the growth spiral is heavily dependent on the presence of impurities. A detailed account of this phenomena may be found in Ref. [5].

2.2 Face Growth Theories

The growth theories discussed so far apply to spherical or quasi spherical particles. However, many crystals (e.g. paracetamol crystals, as shown in Figure 2.10)

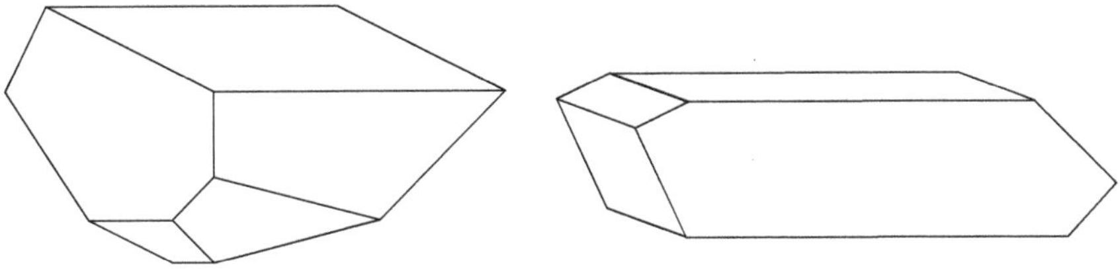

Figure 2.10 Faceted crystal.

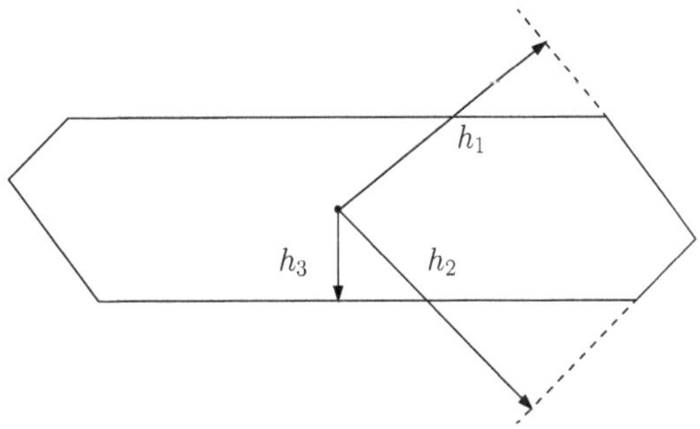

Figure 2.11 A two-dimensional faceted crystal. This crystal is defined by the vector $\mathbf{h} = (h_1 \quad h_2 \quad h_3)^T$.

are 'faceted' with sharp edges and planer faces. Such faceted crystals are often described by a gross shape reference such as 'plate-like', 'needle-like' etc. These references are called crystal habits. For a quantitative treatment, however, we need a precise description of the shape and growth rates for individual faces. We will proceed here with a brief description of this subject and the reader is referred to Ref. [6] for more details.

2.2.1 Shape of a Crystal

A polyhedron is described by a number of attributes and symmetry plays a crucial role in this description. We will limit our discussion to two-dimensional crystals that have either an axis or a centre of symmetry. The crystal shown in Figure 2.11 has a centre of symmetry. The standard method for describing a crystal is to take the symmetry centre as the origin and then draw normals to all the faces. The vector of all such normals describes the crystal.

For a spherical crystal there is only one surface: the surface of the sphere. Growth happens only on that surface and a single growth rate is sufficient. However, for faceted crystals the growth rate of each face is required. These growth rates are usually given by the normal velocity of the faces, dh_i/dt. The growth rate of a face is a function of supersaturation and the crystallographic orientation of the face.

For faceted crystals, both size and shape are important. During growth, a crystal may change its shape along with its size. This situation is shown in

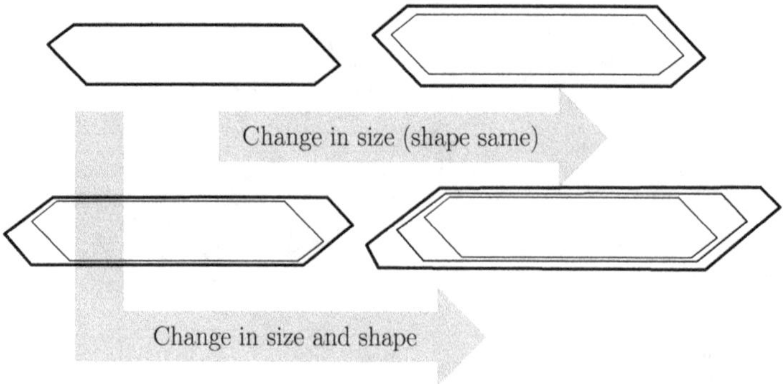

Figure 2.12 Kinetic shape of a crystal. Top: the crystal continues to grow in its kinetic habit. Bottom: the crystal approaches its equilibrium habit.

Figure 2.12. Under one condition, all the faces of a crystal may grow at the same rate and the crystal grows in size by preserving its shape. But as the supersaturation or other growth conditions change, one of the faces will grow at a faster rate than the others. It is clear from the geometry that the faster growing face vanishes from the morphology. This is usually true and faster growing faces are not seen on a grown crystal.

Different crystal faces have different energies. Again, energy minimization can be conducted to obtain the equilibrium shape of a crystal. Although a detail analysis can be conducted, a simplistic argument can be used to reach the same conclusion. Energy minimization dictates that if the surface energy of a face is higher, the area of the face should be less. Complete avoidance of a high energy face may not be possible for geometric reasons and a high energy face should have a minimum possible area. Under equilibrium conditions and flat interfaces this leads to

$$\gamma_i dA_i = \text{constant}.$$

Using geometry, it can be shown that

$$dA_i \propto \frac{1}{h_i},$$

which finally leads to

$$\frac{\gamma_i}{h_i} = \text{constant}.$$

Since the ratio of all h_i can be calculated from surface energies, the equilibrium shape can be obtained.

Under usual conditions, a crystal may not reach this equilibrium shape and crystals are often found to be trapped in a kinetic form. The kinetic shape reached by the crystal is determined by the growth rates of its individual faces.

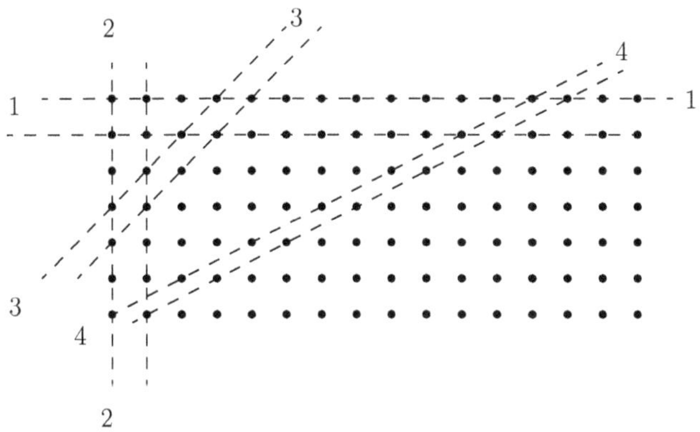

Figure 2.13 Different lattice planes, their interplaner distances, and their growth rates.

Hence, to predict the kinetic shape of a crystal a theory is needed which will predict the growth rate of individual faces.

2.2.2 Laws of Face Growth

The growth rate of individual faces is a function of the crystal lattice, supersaturation, impurities, and solvent. There are several 'laws' that describe the growth rate of various crystal faces.

2.2.2.1 Law of Bravais and Friedel

This law relates the lattice structure of a crystal to its observed shape. According to this law, a family of crystal planes is important (i.e. is seen in a grown crystal) if the interplaner distance is lower. In terms of lattice geometry, higher d_{hkl}, lower importance. This is shown in Figure 2.13.

The Bravais and Friedel law can explain the observed shape of many crystals but fails to explain the relative importance of crystal faces if the glide plane and screw axes are present. To deal with such exceptions, Donnay and Harker proposed a set of additional rules to be applied when the glide plane and screw axis are present. The details of such rules are beyond the scope of this book and the reader is referred to Ref. [7] for details.

2.2.3 Flat, Stepped, and Kinked Faces

A face may be flat, kinked or stepped (see Figure 2.14). Flat faces are compact, grow slowly, and are always observed on the crystal. Stepped faces are less compact and grow at a faster rate. Kink faces are the least compact, grow very fast, and are rarely observed. A theory that can predict whether a crystal surface will be flat stepped or kinked was provided by Hartman and Perdock and is known as periodic bond chain analysis.

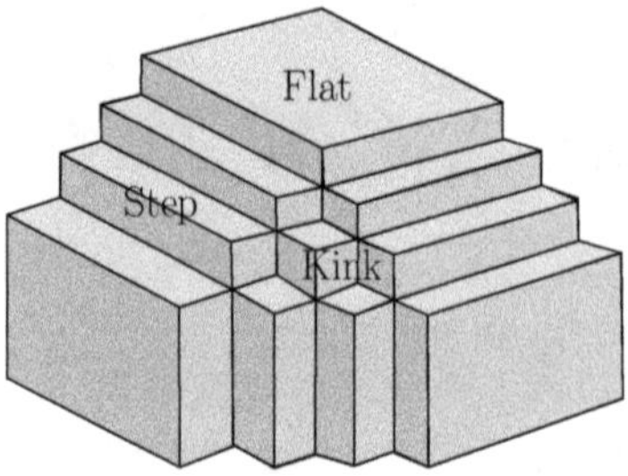

Figure 2.14 Flat, stepped, and kinked crystal faces.

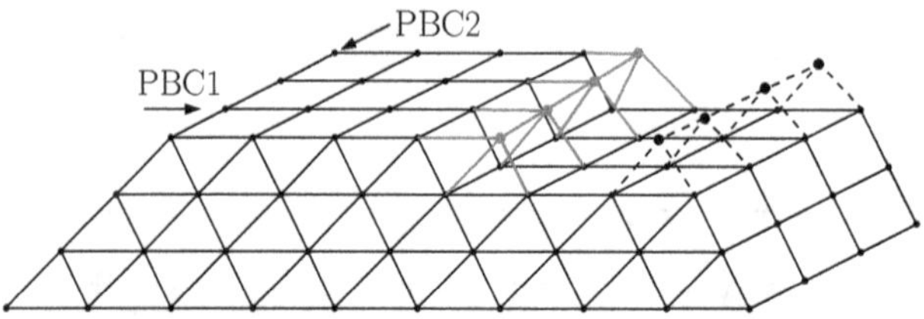

Figure 2.15 The top surface has two periodic bond chains. It grows epitaxially and slowly, and emerges as a flat face. PBC, periodic bond chain.

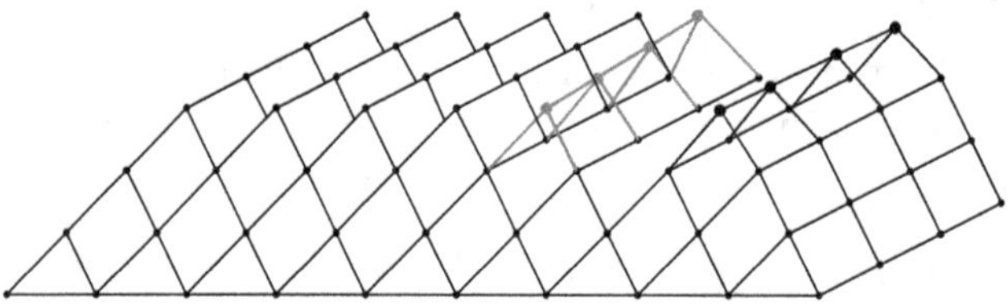

Figure 2.16 The periodic bond chain (PBC) for the lattice shown in Figure 2.15 where one of the bonds is weak.

A periodic bond chain is a chain of strong bonds among the units in the crystal. A crystal with periodic bond chains is shown in Figure 2.15. In this figure the dots represent the growth units and the lines represent strong bonds. It can be seen that for growth of the top surface the growth unit has energetic advantages by depositing on the step (gray dots and lines) rather than elsewhere (black dots and dotted lines).

However, this need not always be the case. Even for the same crystal lattice, if some of the bonds are not strong (as shown in Figure 2.16), there is no particular advantage in depositing on the steps. Hence, individual rows will form and the face will look like steps. If no bond chain exists, individual units may deposit everywhere and the face will be kinked.

These laws provide a general guideline on the relative growth rates of crystal planes in various directions when the lattice structure is known. Usually the lattice structure of a material is known and the relative face growth rates and expected shape can be predicted. However, a quantitative theory that considers industrially relevant conditions has yet to be developed.

2.3 Measurement of Particle Size and Shape

In this section we will discuss various characterization techniques for the determination of particle size and shape. A large number of experimental techniques are available for measuring these two properties. We shall highlight only a few of them. Microscopy and light scattering will be particularly stressed.

The size of a particle is the most important attribute of particulate systems. If particles are bigger than 100 μm, sieving is a good technique for size measurements. This is the simplest and cheapest technique available and yields reasonably accurate results. Sieving data is usually reported as 'mass fraction undersize' or 'mass fraction retained' and can readily be presented as particle size distribution data.

If more precise measurement is needed or particle shape is also important, microscopy should be used. Optical microscopy can usually be employed for particles up to tens of micrometres. For smaller particles, electron microscopy should be employed. However, samples must be absolutely dry for electron microscopy and hence artifacts may be introduced when slurry/colloids are dried for electron microscopy. Recently, environmental SEM has been introduced where such restrictions are somewhat relaxed [8].

To obtain particle size distribution using microscopy, good quality images of a large number of particles must be obtained. Such images are then analyzed using suitable image processing techniques to obtain the particle size and shape distribution. A number of software packages (Matlab and Image J to name a couple) are available for image processing. Since particles are non-spherical for most practical applications, a number of size and shape descriptors are used to quantify particle size and shape. A brief list of such size–shape indicators is given below:

- Projected area diameter: The diameter of the circle having the same area as the projected area of the particles.
- Feret's diameter: The distance between two tangents on opposite sides of the particle, parallel to some fixed direction.
- Martin's diameter: The length of the line that divides the image area into equal halves. The line may be in any direction, but should be consistent for all particles.

Apart from microscopy there is a variety of techniques that can be used for measurement of size. For example, gravity settling or centrifugation can be used for segregation of particles into various size classes of narrow range and subsequent measurements. Another interesting technique for measurement of particle size distribution in a dilute suspension is the electric pulse counter. In this instrument, the suspension is passed through an orifice. The suspension should be very dilute so that only one particle can pass through the orifice at a time. Each time a particle passes through the orifice, an electric pulse is generated in a pair of electrodes mounted on either side of the orifice. The magnitude of the pulse is proportional to the cross-sectional area of the particle. These pulses can be recorded and converted into particle size distribution.

As evident from the foregoing discussion, accurate measurement of size and shape using microscopy is a rigorous process. With optical microscopy, such rigour yields accurate results, but for electron microscopy, the sample must be dry and free of any component that may melt or evaporate under ultra high vacuum. This feature of electron microscopy requires particles to be cleaned of all such components and in many cases this leads to particle aggregation. Hence, the particles observed under the electron microscope may not be the same as the particles in the suspension. Alternative techniques are therefore required for measurement of the size and shape of submicron particles.

Light scattering is a very popular technique for measuring particle size where a diluted dispersion is used as the sample. Occasionally, it is used for measuring particle shape as well. Although no sample preparation step is involved in this case other than diluting and filtering the colloid for dust, the interpretation of the data is not straightforward and the presence of a small amount of large particle impurity can spoil the measurement.

Usually, neither microscopy nor light scattering is a foolproof technique for measuring the size and shape of submicron particles. A battery of tests is usually carried out to obtain the size and shape of small particles.

2.3.1 Optical Microscopy

An optical microscope provides an enlarged digital image of an object. Usually illumination of the object is done using white light. An objective lens and an eye-piece are used for enlargement. The main components of a microscope are the illumination system, the objective lenses, and the eyepiece. For transparent biological samples transmitted light is mostly used. For non-transparent samples a reflected light mode is used in which the light is directed from the objective lens itself. The objective lens makes a real image of the object and magnifies it.

Objects up to 1 μm in size can be observed clearly using a 100X objective in an optical microscope. For smaller particles, the resolving power of the microscope becomes a limiting factor and although the object can be viewed, its image becomes blurred. If the distance between two points is comparable with the wavelength of the light, the light that comes from different parts of the object interfere and hence a sharp image cannot be produced. It can be shown using geometric optics that [9]

$$d_p \approx \frac{\lambda}{n_0 \sin\theta}, \tag{2.5}$$

where d_p is the resolving power, λ is the wavelength of light, n_0 is the refractive index (RI) of the medium, and θ is the view angle. It is clear that if the RI of the medium is higher, d_p is lower. For this reason, oil is often used to fill the space between the sample and the objective lens to achieve higher resolving power. Even using oil, d_p below 0.2 μm is very difficult.

It may be noted that although the edges of a particle will be blurred, the presence of a particle smaller than 0.2 μm can be detected. In some cases, the shape or size of a particle is not very important. Rather, the number of particles present in a sample is important. In that case, optical microscopy can be used for particles around 0.1 μm. The instrument that has been specifically developed for this task is called an ultramicroscope. In an ultramicroscope, the light source and the objective lenses are kept at right angles. The light that is scattered by the particles is collected by the objective lens. Hence, the particles appear as bright spots on a dark background.

2.3.2 Electron Microscopy

To improve the resolving power significantly, the wavelength of the 'light' must be decreased (see eqn 2.5). Hence, instead of visible light, a suitable radiation with a small enough wavelength should be used. One such readily available radiation is the flow of electrons from a heated filament. Electrons have a wavelength of 0.025 Å at 200 kV accelerating voltage and hence the corresponding resolving power is in the order of angstroms.

In optical microscopy, the refraction of light (i.e. the bending of its path) is used to form an enlarged image. The same technique is used for electron microscopy, but instead of glass lenses, an electro-magnetic field is used through conducting coils. The image is observed by electronics and is not visible to human eye. The details of the imaging are dependent on the type of electron microscopy used. Of the many electron microscopy techniques available, two are very important for the measurement of particle shape and size:

- Transmission electron microscopy (TEM): The electron beam transmits through the sample and an image is captured. In this case essentially the projection of the object on a horizontal plane is obtained.
- Scanning electron microscopy (SEM): An electron gun sprays electrons to the sample and the backscattered electrons are captured. The contrast is created on the basis of the depth from which the electrons were scattered. Thereby, a sense of depth is created in the image.

2.3.3 Light Scattering

Light is electromagnetic radiation. When it encounters another medium (particle) while passing through the solvent, absorption and scattering occur [10].

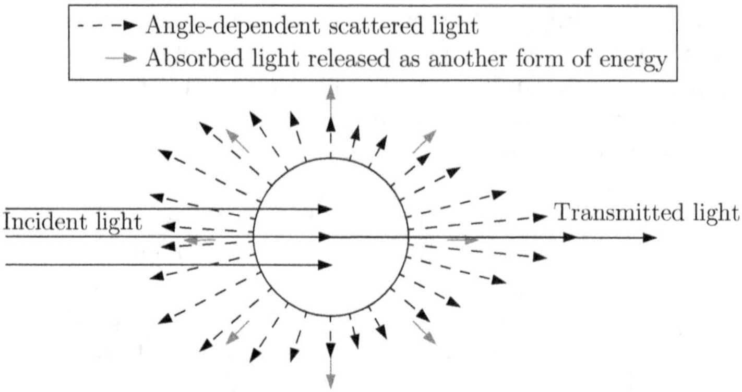

Figure 2.17 Scattering of light from a particle.

The unabsorbed and unscattered portion transmits through the colloid. A part of the electromagnetic radiation that is taken up by the particle is transferred into another form of energy and the rest is radiated back in all directions as light. The portion that is transferred into another form is said to be absorbed while the portion that is radiated back in all directions is called scattered. Two important properties of the scattered light are (i) the frequency of light remains unchanged during the scattering process and (ii) the intensity of the scattered light from a particle is a function of the angle. This is illustrated in Figure 2.17.

The nature of the scattering depends strongly on the size of the scatterer (particle) with respect to the wavelength of the light. Hence, the theory of light scattering is divided into three different categories:

- Rayleigh scattering: The size of the scattering centres is much less than the wavelength of the radiation and particles do not absorb light. Applicable to non-absorbing particles that are much smaller than the wavelength of the radiation, e.g. polymer molecules.
- Debye scattering: Applies when the particle size is comparable to the wavelength of light.
- Mie scattering: Small particles simultaneously absorb and scatter light.

In this chapter we shall limit ourselves to studying Rayleigh scattering, keeping in mind the determination of the size of polymer or protein molecules.

2.3.3.1 Rayleigh Scattering

Rayleigh scattering is the basic scattering theory and we will discuss it here in some detail [11]. Let us consider the case of polymer/protein molecules dissolved in a solvent. The molecules are usually about a nanometre in size and hence are much smaller than the wavelength of visible light (usually 530 nm

Figure 2.18 Rayleigh scattering.

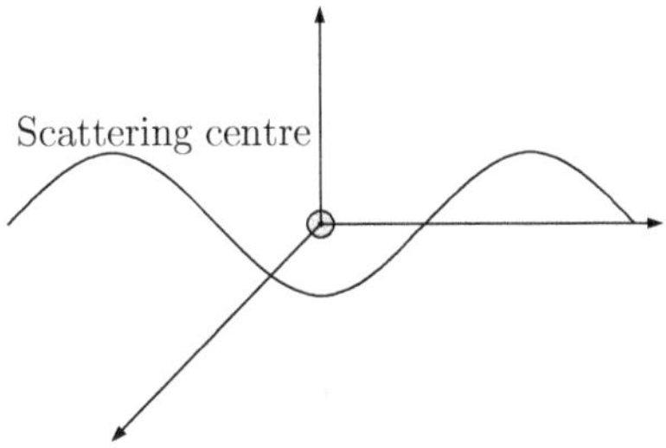

green LASER is used in light-scattering experiments). Polymer molecules scatter a reasonable amount of light for effective detection. Solvent molecules may also scatter light, but because they are densely spaced and randomly oriented, the scattered radiations from solvent molecules undergo destructive interference. For this reason, a pure solvent usually remains a very weak scatterer and we shall consider scattering by the polymer molecules (particles) only.

Light is an electromagnetic radiation. Hence, if we consider only the electric part, like all other waves it is an oscillating electric field that varies over space and time:

$$\mathbf{E} = \mathbf{E}_0 \cos\left[2\pi\left(\nu t - \frac{x}{\lambda}\right)\right].$$

For light-scattering experiments polarized laser light is usually used and hence the situation can be depicted as shown in Figure 2.18. The wave lies in the plane of the paper. The electromagnetic radiation is scattered by a scattering centre, shown at the origin. In Rayleigh scattering the scattering centre is much smaller than λ and hence the electric field is same everywhere in the scatterer.

The scatterer molecule may be neutral, but as discussed in section 3.1, even a neutral molecule develops an induced dipole moment when placed in an electric field. The dipole moment of this induced dipole is

$$\mu = \alpha\mathbf{E} = \alpha\mathbf{E}_0 \cos\left[2\pi\left(\nu t - \frac{x}{\lambda}\right)\right],$$

where α is a material property called the polarizability of the scatterer and x is the distance of the scatterer from the origin. Since we are free to choose the coordinate system, we can place the scatterer at the origin to simplify the situation. With this choice of origin, the above equation becomes

$$\mu = \alpha\mathbf{E} = \alpha\mathbf{E}_0 \cos(2\pi\nu t).$$

It can be seen from this equation that the induced dipole moment oscillates with the frequency of the light. How can the dipole moment change with time? Either the charges change by keeping the distance between them constant or the distance changes with time. Either view is possible. Let us consider the second one, i.e. oscillation of two charges of constant value. With this picture in mind, the above equation may be considered as a charge $\alpha E_0/l$ separated by a time varying distance $l\cos(2\pi\nu t)$. The length l is an unknown quantity but it will cancel. The acceleration of the pair of charges therefore is

$$\mathbf{a} = -4l\pi^2\nu^2 \cos(2\pi\nu t).$$

Now, according to classical electrodynamics, accelerating charges emit radiation in all directions. The electric field $\mathbf{E_s}$ at a distance r from an accelerating charge q can be obtained by solving Maxwell's equation and is given by [12]

$$\mathbf{E_s} = \frac{q\mathbf{a}\sin\phi_z}{4\pi\epsilon c^2\mathbf{r}}.$$

Here ϕ_z is the angular orientation of the observer (see Figure 2.17) and $\epsilon = \epsilon_0\epsilon_r$ is the permittivity of the medium. Using this equation, the scattered electric field is given by

$$\mathbf{E_s} = -\frac{\pi\nu^2\alpha\mathbf{E_0}\cos(2\pi\nu t)\sin\phi_z}{\epsilon c^2 r}.$$

Note that the length variable l cancels at this stage. Because the intensity of a field is proportional to the square of its strength, the intensity of scattered light at a given position is

$$i_s \propto \frac{\pi^2\nu^4\alpha^2\mathbf{E_0}^2\cos^2(2\pi\nu t)\sin^2\phi_z}{\epsilon^2c^4r^2}.$$

Because the intensity of the incident beam, i_0, can be written as $i_0 \propto \mathbf{E_0}^2\cos^2(2\pi\nu t)$, the intensity ratio is given by

$$\frac{i_s}{i_0} = \frac{\pi^2\nu^4\alpha^2\sin^2\phi_z}{\epsilon^2c^4r^2}.$$

The intensity ratio can be measured using a photometer, but how does it relate to the size of the scatterer? The link is through polarizability. The larger the molecule, the more polarizable it is. It can be shown that the polarizability of a molecule can be written as (the Clausius–Mosotti equation)

$$\alpha = \frac{3M\epsilon_0}{\rho N_A}\frac{n^2-1}{n^2+2},$$

where M is the molecular weight of the scatterer and n is the RI of the molecule. For particles which are not a single molecule, the concept of polarizability still works. It can be shown that the polarizability is proportional to particle volume and hence the intensity ratio is proportional to R^6: a very strong dependance. For this reason, the scattering by small particles often becomes negligible with respect to large particles and it is very difficult to detect small particles in the presence of large particles using light scattering.

2.3.3.2 Static and Dynamic Light-Scattering Techniques

In static light scattering, the time average intensity of the scattered light is measured and the particle size is determined using a technique similar to that shown above (for details see Ref. [12]). In dynamic light scattering, the transient scattered light intensity is measured instead of its time average. Specialized hardware is needed for this measurement. The transient in the scattered light is because of the movement of the particle. Although the particle scattered light of the same wavelength as that of incident light, the wavelength that is observed by the stationary detector is slightly different from that because of the diffusive movement of the particles (Doppler effect). The particle diffusivity or diffusive velocity can therefore be measured from this dynamic data. The dynamic data is measured and used in terms of an auto correlation function. The co-relation between the subsequent intensities decreases as the time delay between them increases. This auto correlation function can be related to particle diffusivity. Hence, the particle diffusivity can be obtained. From the particle diffusivity, the size can be obtained (see section 4.5.1).

2.4 Exercises

Exercises 2.1 Uniform copper micro particles can be grown by seeded growth of copper nanoparticles. The copper nanoparticle seeds range from a few nanometres to 200 nm. The copper ions needed for growth are supplied by reducing copper acetate using hydrazine hydrate. Formulate a model to predict the effect of hydrazine hydrate concentration on the polydispersity of the final particles.

Exercises 2.2 It is planned to produce elongated nanorods of zinc oxide by hydrolysis of zinc nitrate in water under basic conditions. The face whose growth needs to be blocked is a stepped face. What are the considerations needed for selection of the surfactant? Will the solvent have any effect on the process?

Exercises 2.3 If a face disappears from the crystal surface, will the h corresponding to that face remain constant? The face may reappear later.

Exercises 2.4 Consider an initial normally distributed set of particles growing under diffusion controlled growth. Obtain an expression for the evolution of the coefficient of variance as a function of time.

Bibliography

1 A. E. Nielsen. *Kinetics of Precipitation*. Pergamon Press, Oxford, 1964.

2 R. B. Bird, W. E. Stewart, and E. N. Lightfoot. *Transport Phenomenona*, 2nd edition. Wiley, 2007.

3 A. J. Malkin, Y. G. Kuznetsov, and A. McPherson. In situ atomic force microscopy studies of surface morphology, growth kinetics, defect structure and dissolution in macromolecular crystallization. *Journal of Crystal Growth*, *196*(2–4):471–488, 1999.

4 X. Peng, J. Wickham, and A. P. Alivisatos. Kinetics of II–VI and III–V colloidal semiconductor nanocrystal growth: focusing of size distributions. *Journal of the American Chemical Society*, *120*(21):5343–5344, 1998.

5 K. Sangwal. *Additives and Crystallization Processes: From Fundamentals to Applications*. John Wiley & Sons, 2007.

6 A. S. Myerson. *Handbook of Industrial Crystallization*, 2nd edition. Butterworth-Heinemann, 2002.

7 P. Hartman. *Crystal Growth: An Introduction*. North Holland, 1973.

8 D. Stokes. *Principles and Practice of Variable Pressure/Environmental Scanning Electron Microscopy (VP-ESEM)*. John Wiley & Sons, 2008.

9 D. B. Murphy. *Fundamentals of Light Microscopy and Electronic Imaging*. John Wiley & Sons, 2002.

10 C. F. Bohren and D. R. Huffman. *Absorption and Scattering of Light by Small Particles*. Wiley, 1983.

11 R. J. Hunter. *Foundations of Colloid Science*, 2nd edition. Oxford University Press, 2001.

12 P. C. Hiemenz. *Principles of Colloid and Surface Chemistry*, 2nd edition. Marcel Dekker, 1986.

3

Inter-Particle Forces

Particles created by nucleation and growth are not in a stable equilibrium due to the existence of a large surface area. The system can reduce its surface energy by aggregation of particles. In many cases, large surface area is a favorable property of a particulate system and aggregation is an undesirable process. For others, separation of particles is needed and aggregation is a desirable process.

Aggregation occurs due to two distinct processes:

- movement of particles: particles move and collide with each other in a solution by Brownian or flow-driven motion
- attachment: attractive forces between particles hold together the colliding pairs to form aggregates.

Sometimes aggregation and growth occur simultaneously and in such cases growth also helps in cementing the aggregating pairs by depositing new material on the assembly. In this chapter we shall discuss the nature of force between particles.

This force is generally attractive and the origin of the attractive force between particles is the attractive force between atoms and molecules. Hence, we shall start with a brief discussion of various types of attractive forces between atoms and molecules, and then discuss how these forces lead to inter-particle attraction.

Two aspects of the inter-atomic and inter-particle force are of importance: first, the magnitude of these forces with respect to the magnitude of thermal fluctuation, kT, and the distance dependance. Usually, the attractive force decays with distance. If it decays very slowly with distance, it can influence a particle far away. Such types of forces have different consequences to forces which are active only at close proximity.

Another important point to note is the relation between force and energy. If two particles experience a distance-dependant force of $F(r)$, then the interaction energy of these two particles at a distance r is given by

$$w(r) = -\int_{\infty}^{r} F(r')dr'.$$

Engineering of Submicron Particles: Fundamental Concepts and Models, First Edition. Jayanta Chakraborty.

Companion website: http://booksupport.wiley.com

In some situations force description may be advantageous but energy description facilitates calculations in many cases because energy is a scaler quantity. In our analysis we shall switch from one description to the other depending on the situation. Let us start by classifying inter-molecular forces.

3.1 Inter-Molecular Forces

One of the common molecular species in an aqueous solution is ions. Dipoles are also common. The water molecule is an example of a dipolar molecule. All such species will interact through electrostatic interaction and can be described by classical electrodynamics. Many molecules do not have a permanent dipole moment. However, in presence of an electric field, the electron cloud deforms and produces an 'induced' dipole. The electric field may be due to another permanent dipolar molecule. Some of the inter-molecular interactions discussed in this text are:

- ion–ion interactions
- ion–dipole interactions
- dipole–dipole interactions
- dipole–induced dipole interactions
- induced dipole–induced dipole interactions: when the inducing element itself is an induced dipole. This force is quantum mechanical in origin.

There are many other short- and long-range interactions that are also very important. The reader is referred to advanced texts [1] for a more elaborate description.

Most of these interaction energies can be calculated using classical electrodynamics but some (e.g. induced dipole–induced dipole interactions) are quantum mechanical in origin and cannot be explained from classical electrodynamics. Let us start by calculating the interaction energy between two ions: charge–charge interactions.

3.1.1 Charge–Charge Interactions

The electric field at a distance r from a charge Q is given by

$$E_1 = \frac{Q_1}{4\pi\epsilon_0\epsilon_r r^2}.$$

Hence, for two atoms having charge Q_1 and Q_2, respectively, the force between them is

$$F(r) = \frac{Q_1 Q_2}{4\pi\epsilon_0\epsilon_r r^2}.$$

Hence, the interaction energy at a distance r is

$$w(r) = -\int_{\infty}^{r} \frac{Q_1 Q_2}{4\pi\epsilon_0\epsilon_r r^2} dr = \frac{Q_1 Q_2}{4\pi\epsilon_0\epsilon_r r}.$$

The constant ϵ_0 is the dielectric permittivity of vacuum and ϵ_r is the relative permittivity of a medium.

Example 3.1 Calculate the interaction energy between Na^+ and Cl^- ions in NaCl crystal in a vacuum. The value of ϵ_0 is 8.854×10^{-12} in SI units.

Solution: The charge on each of the ions corresponds to one electronic charge, i.e. 1.602×10^{-19} Q. The ions are in atomic contact in the crystal, hence the centre-to-centre distance is the sum of their atomic radii. The atomic radii of sodium and chlorine are 190 pm and 79 pm, respectively. Hence, the interaction energy is

$$w(r) = -\frac{(1.602 \times 10^{-19})^2}{4\pi 8.854 \times 10^{-12} \times 269 \times 10^{-12}} = -8.4 \times 10^{-19}\text{J} \approx -200kT.$$

□

It can be seen that the ions are held by a very strong force and hence this crystal must be very stable. However, the crystal dissolves in water because water has very high dielectric constant, $\epsilon_r = 78$. The interaction energy in water is therefore $(200/78)kT = 2.5kT$. Hence, in this case the thermal fluctuation energy will be able to dissociate the crystal. Another point to note here is the distance dependance of this interaction energy. It can be seen that the force decays as $1/r$, which is much weaker than other forces (usually varies as $1/r^6$). Hence an ion–ion interaction is considered to be a long-range force.

3.1.2 Charge–Dipole Interactions

Ions usually exist in polar solvents. The most common type of polar molecule is dipoles, for example water. The interaction of water with ions like sodium or chloride is governed by charge–dipole interactions. The geometry of the system is shown in Figure 3.1. The interaction energy can be obtained by a method similar to that used for charge–charge interactions and is given by

$$w(r, \theta) = -\frac{Qu\cos\theta}{4\pi\epsilon_0\epsilon_r r^2},$$

where u is the dipole moment of the polar molecule. In this equation it is assumed that $r >> l$. A slightly different form is obtained without this assumption. It can be seen that the interaction energy is dependent on the orientation of the dipole and decays at a much faster rate than for charge–charge interactions.

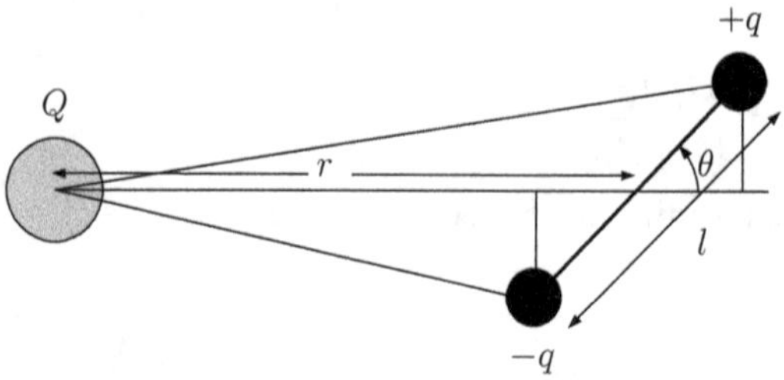

Figure 3.1 Charge–dipole interaction.

Example 3.2 Calculate the interaction energy between Na+ and water molecules when $r = 0.235$ nm (in contact). The dipole moment of water is 1.85 Debye.

Solution: 1 Debye = 3.336×10^{-30} Qm. Hence, in a vacuum, in contact, and taking $\theta = 0$,

$$w(r,0) = -\frac{1.602\times10^{-19}\times1.85\times3.336\times10^{-30}}{4\pi\times8.854\times10^{-12}\times(0.235\times10^{-9})^2} = -1.6\times10^{-19}\,\text{J} = -40\,kT. \quad \square$$

It can be seen that the interaction energy is reasonably high, although not as high as the charge–charge interaction. Now, what is the interaction energy if the centre-to-centre distance is 1 nm instead of 0.235 nm? The interaction energy is calculated to be only $2kT$ in this case. Such a small amount is not enough to hold the dipoles at a stationary angular orientation and the dipole starts to rotate with respect to the ion due to thermal fluctuation at such distances. Hence the interaction energy for this case will be the angular average of all possible orientations. However, all orientations are not equally likely. The probability of a particular angular orientation is a function of its energy. Specifically, their relative frequency will follow a Boltzmann distribution. Hence, the average energy of a freely rotating dipole is obtained as

$$\overline{w}(r) = \frac{\int w(r,\theta)\exp(-w(r,\theta)/kT) sin(\theta) d\theta}{\int \exp(-w(r,\theta)/kT)\sin(\theta)d\theta}.$$

The above integration leads to

$$\overline{w}(r) = -\frac{Q^2u^2}{6(4\pi\epsilon_0\epsilon)^2kTr^4}.$$

It can be seen that the distance dependance of this angle-averaged interaction is much stronger than the fixed angle case.

3.1.3 Dipole–Dipole Interactions

Another frequently encountered interaction is dipole–dipole interaction. This situation is shown in the Figure 3.2. Note that the dipole at the right is not on

Figure 3.2 Dipole–dipole interaction.

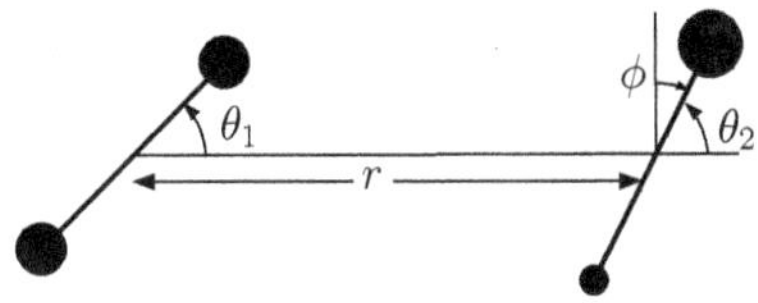

the plane of paper. The angle θ_2 is the angle its projection makes on the plane of paper and ϕ is the angle between the projection and the dipole. Here also the dipoles can be either fixed or rotating depending on the strength of the interaction, as discussed previously. For fixed dipoles, the interaction energy can be calculated as

$$w(r, \theta_1, \theta_2, \phi) = -\frac{u_1 u_2[2\cos\theta_1 \cos\theta_2 - \sin\theta_1 \sin\theta_2 \cos\phi]}{4\pi\epsilon_0\epsilon_r r^3}.$$

If both the dipoles rotate freely, an angle average should be obtained and the interaction energy becomes

$$\overline{w}(r) = -\frac{u_1^2 u_2^2}{3[4\pi\epsilon_0\epsilon]^2 kTr^6}.$$

This interaction is called the *Keesom interaction* or *orientation interaction* and it is one component of the famous van der Waals interaction.

3.1.4 Dipole–Induced Dipole Interactions

Not all molecules have a permanent dipole. However, if they are in an electric field, the electron cloud of the molecule becomes polarized and the molecule obtains an induced dipole. Under an identical electric field, different molecules polarize to different extents. The material property that determines the degree of polarization possible is called polarizability and is denoted by α. The induced dipole moment is proportional to the applied electric field and is given by $u_{ind} = \alpha E$. For the present case, the electric field is due to the dipole moment of the permanent dipole present.

The angle-averaged interaction energy for this case can also be calculated in a similar manner and is given by

$$\overline{w}(r) = -\frac{u^2\alpha_0}{(4\pi\epsilon_0\epsilon)^2 r^6}.$$

This interaction is called the *Debye interaction* or *induction interaction* and it is another component of the van der Waals force.

Example 3.3 Estimate the induced dipole moment in a methane molecule from an isolated sodium ion located at a distance of 0.5 nm from its centre. The medium is a vacuum. Compare the induced dipole moment with the dipole moment of the water molecule.

Solution: The electric field at a distance r is given by

$$E(r) = \frac{e}{4\pi\epsilon_0 r^2} = \frac{1.6 \times 10^{-19}}{4 \times 3.142 \times 8.85 \times 10^{-12} \times (0.5 \times 10^{-9})^2}$$
$$= 5.75 \times 10^9.$$

Therefore, the induced dipole moment is

$$u_{ind} = \alpha_0 E = 2.88 \times 10^{-40} \times 5.75 \times 10^9 = 1.65 \times 10^{-30}\text{Qm}$$
$$= 0.496 \text{ Debye}$$

□

3.1.5 Induced Dipole–Induced Dipole Interactions

A strange interaction happens between a pair of molecules having no permanent dipole moment. The angle-averaged interaction energy for this case is given by

$$\overline{w}(r) = -\frac{3}{2}\frac{I_1 I_2}{I_1 + I_2}\frac{\alpha_{01}\alpha_{02}}{[4\pi\epsilon_0]^2 r^6}.$$

This interaction energy is called *the London interaction* energy or *dispersion interaction* energy. This is the strongest of the three components of the van der Waals interactions and it is present in all molecules. It can be seen that the ionization potentials (I_1 and I_2) and polarizability (α_{01} and α_{02}) are two important material properties for this case.

3.1.6 van der Waals Interaction

It can be seen that all three components of the van der Waals interaction energy are attractive and have the same distance dependance, i.e. $1/r^6$. Hence, the van der Waals attraction between two molecules can be written as

$$w_{vdw}(r) = -\frac{C_{disp}}{r^6} - \frac{C_{ind}}{r^6} - \frac{C_{orient}}{r^6} = -\frac{C_{vdw}}{r^6}.$$

The coefficients can be calculated and are given for a few cases in Table 3.1.

It can be seen that the dispersion interaction is the major component of van der Waals forces and this component exists between all kind of molecules.

Table 3.1 Relative magnitude of various components of van der Waals forces in 10^{-79} J m^6

Interacting units	C_{ind}	C_{orient}	C_{disp}	Total	Experimental
Ne/Ne	0	0	4	4	4
CH_4/CH_4	0	0	102	102	101
HCl/HCl	6	11	106	123	123
NH_3/NH_3	10	38	63	111	162

Values obtained from Table 6.1 of Ref. [1].

Example 3.4 Calculate the van der Waals interaction energy between two methane molecules (a) when in contact and (b) when 1 nm away.

Solution: The interaction energy is

$$w(r = 0.4\text{nm}) = -\frac{102 \times 10^{-79}}{(4 \times 10^{-10})^6}\text{J} = 2.5 \times 10^{-21}\text{J} = 0.6\,kT$$

$$w(r = 1\text{nm}) = -\frac{102 \times 10^{-79}}{(1 \times 10^{-9})^6}\text{J} = 1.02 \times 10^{-23}\text{J} = 0.002\,kT.$$

□

Example 3.5 The *self energy* of a molecule is defined as the energy required to place the molecule in its position by bringing it from infinite distance while all other molecules remain fixed at their respective positions. Calculate the self energy of a molecule assuming that the interaction energy can be taken as pair-wise additive.

Solution: If the density of the material in terms of number of molecules/unit volume is denoted by ρ, the self energy (μ) can be written as

$$\mu = \int_0^\infty w(r)\rho 4\pi r^2 dr = -\frac{4\pi C\rho}{3\sigma^3}.$$

□

3.1.7 Repulsive Potential and the Net Interaction Energy

When two molecules come very close together, their electron clouds start to overlap and the molecules start to repel each other instead of attracting. The repulsive potential at contact is known as Born repulsion. The distance dependance of this repulsive force is much stronger than the attractive part and is given by $1/r^{12}$. The overall interaction potential is therefore given by

$$w(r) = -\frac{C_{vdw}}{r^6} + \frac{C_B}{r^{12}}.$$

This is known as the Leonard–Jones potential or 12–6 potential. Although other variations of inter-molecular potential are in use, this is the most widely used interaction potential.

3.2 Inter-Particle Forces

Particles are made by molecules. The number of molecules in a particle may range from a few thousands to a few million or more. The interaction energy between the particles may be viewed as the energy necessary to build the molecular arrangement comprising the two particles minus the energy necessary to build the individual particles. Hence, this interaction energy may be computed in the following way:

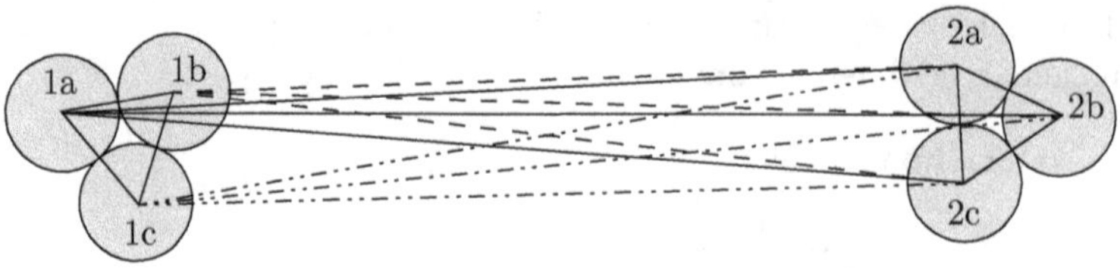

Figure 3.3 Hamaker's pairwise additivity approach.

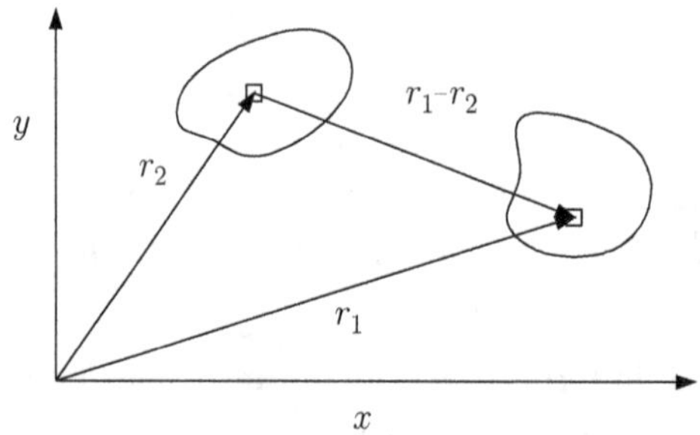

Figure 3.4 Hamaker's pairwise additivity approach.

- Place the first particle in space.
- Bring each of the molecules of the second particle one by one and compute its interaction energy with each of the molecules of the first particle.
- Add all the interaction energies to obtain the net energy of interaction between the two particles.

This situation is shown in Figure 3.3. In this case, the total interaction energy between the two particles is the sum of the following molecular interactions: (1a–2a, 1b–2a, 1c–2a; 1a–2b, 1b–2b, 1c–2b; 1a–2c, 1b–2c, 1c–2c). This approach was first studied by Hamaker in 1937 and is known as Hamaker's pairwise additivity approach. In a practical situation, such molecular level additions are not possible and a continuum approximation is taken, as discussed next.

3.2.1 Hamaker's Pairwise Additivity Approach

Let us consider two bodies of arbitrary shape, as shown in Figure 3.4. Usually, particles contain a large number of molecules and the distances between particles are at least an order of magnitude larger than the molecular dimensions. Hence, the addition of energies is conducted by taking a continuum-like approach instead of counting individual pairs. Let us consider a pair of elementary volumes (dv_1 and dv_2) inside each of the particles, as shown. Because the distances are large compared to the molecular dimensions, each molecule in the small volume in particle 1 may be considered equidistant from the molecules in elementary volume 2. If the position vectors of elementary volumes 1 and 2 are $\mathbf{r}_1$ and $\mathbf{r}_2$, respectively, the distance between them is given by $\mathbf{r} = \mathbf{r}_1 - \mathbf{r}_2$.

The energy of interaction between a molecule residing in v_1 and another residing in v_2 is given by

$$w_{12}(r) = \frac{C_{12}}{|\mathbf{r_1} - \mathbf{r_2}|^6}.$$

Let us denote the number of molecules per unit volume by ρ_1 and ρ_2. Since dv_2 contains $\rho_2 dv_2$ molecules, for a single molecule in v_1 the interaction energy is

$$d\Phi_{12}(r) = w_{12}(r)\rho_2 dv_2.$$

Since dv_1 contains $\rho_1 dv_1$ molecules, the total interaction energy between these two elementary volumes is given by

$$d\Phi_T(r) = w_{12}(r)\rho_2 dv_2 \rho_1 dv_1.$$

The total interaction energy between these two bodies is obtained as:

$$\Phi_T = \int_{B1}\int_{B2} w_{12}(r)\rho_1\rho_2 dv_1 dv_2.$$

Although the integration is very complicated for arbitrary shapes, it becomes significantly simplified for regular geometries, as illustrated next.

Example 3.6 Obtain the total interaction energy between a molecule and a semi-infinite solid. The molecular densities of the two bodies are ρ_1 and ρ_2 respectively.

Solution: The geometry is shown in Figure 3.5.

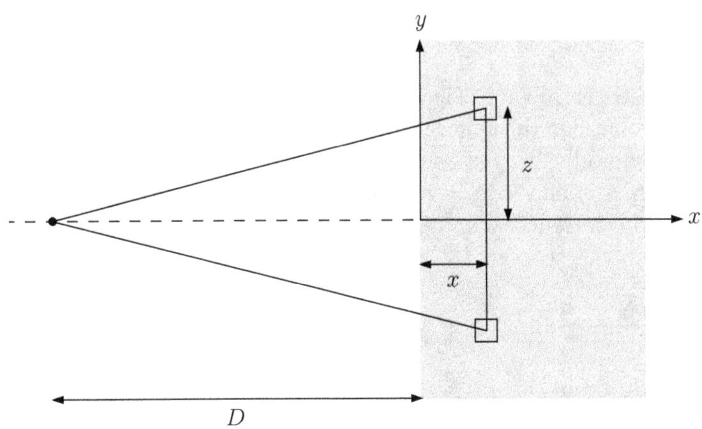

Figure 3.5 Molecule–semi-infinite body interaction.

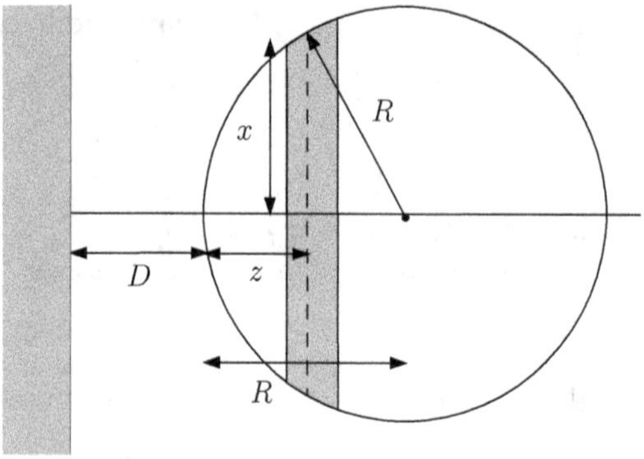

Figure 3.6 Sphere–semi-infinite body interaction.

At a depth x, the distance of the circular ring from the molecule is $\sqrt{(D+x)^2+z^2}$. The elementary volume is $2\pi z dz dx$. Hence, the total interaction energy is

$$-\int_0^{\infty} dx \int_0^{\infty} dz \frac{2C\rho\pi z}{((D+x)^2+z^2)^3},$$

which yields (using Mathematica)

$$-\frac{C\rho\pi}{6D^3}.$$

□

The important point to note here is that the distance dependance of the interaction energy is different to that of the molecular interaction.

Example 3.7 Obtain the total interaction energy between a sphere and a semi-infinite solid.

Solution: The geometry is shown in Figure 3.6.

The elementary volume in the sphere is the small strip that is at a fixed distance $D+z$ from the body. The radius of the disc is given by chord theorem, as shown in section A.3, and is $x^2 = z(2R-z)$. The number of molecules is therefore $\pi z(2R-z)\rho dz$. Hence, according to the previous derivation, the total interaction energy for this strip is

$$-\frac{C\rho_1\rho_2\pi^2 z(2R-z)dz}{6(D+z)^3}.$$

For the entire sphere, it must be integrated:

$$\phi_T = -\int_0^{2R} \frac{C\rho_1\rho_2\pi^2 z(2R-z)dz}{6(D+z)^3}.$$

Table 3.2 van der Waals interactions among various macroscopic bodies

Geometry	Van der Waals interaction: Energy	Van der Waals interaction: Force
Two flat surfaces (per unit area)	$-\frac{A}{12\pi D^2}$	$-\frac{A}{6\pi D^3}$
Two spheres of radii R_1 and R_2	$-\frac{A}{6D}\left(\frac{R_1R_2}{R_1+R_2}\right)$	$-\frac{A}{6D^2}\left(\frac{R_1R_2}{R_1+R_2}\right)$
Sphere of radius R near a semi-infinite surface	$-\frac{AR}{6D}$	$-\frac{AR}{6D^2}$
Two cylinders of radii R_1 and R_2 crossed at 90°	$\frac{-A\sqrt{R_1R_2}}{6D}$	$-\frac{A\sqrt{R_1R_2}}{6D^2}$

The constant $C\rho_1\rho_2\pi^2$ is called Hamaker's constant and usually denoted by A. Hence, the above expression becomes

$$\phi_T = -\frac{A}{6}\int_0^{2R}\frac{z(2R-z)dz}{(D+z)^3}.$$

This gives

$$\phi_T = -\frac{RA}{3}\frac{2R(D+R)+D(D+2R)\log\left(\frac{D}{D+2R}\right)}{2DR(D+2R)}.$$

For the case $D << R$, this reduces to

$$\phi_T = -\frac{AR}{3}\left(\frac{R+D\log(D/2R)}{2DR}\right).$$

The second term in the numerator is of the order of D^2/R and hence can be removed. Hence, for separation much less than the particle diameter,

$$\phi_T = -\frac{AR}{6D}.$$

□

A similar method can be used for various geometries and a few important results are summarized in Table 3.2.

3.2.2 Lifshitz's Theory

Hamaker's theory does not take into account the presence of a medium between the particles, but we have seen previously that the presence of medium changes the van der Waals interaction. Hence the inter-particle interaction should also depend on the medium in which the particles are suspended. Lifshitz corrected this limitation of Hamaker's theory and calculated the Hamaker constants (shown in Table 3.3) between various pairs of objects

Table 3.3 Hamaker constants calculated from Lifshitz's theory

1	3	2	Hamaker constant A (10^{-20} J)
Air	Water	Air	3.7
PTFE	Water	PTFE	0.29
Water	Hydrocarbon	Water	0.3–0.5
Mica	Water	Mica	2.0
α-Alumina	Water	α-Alumina	4.2
Ag, Au, Cu	Water	Ag, Au, Cu	10–40
Silica	Dodecane	Silica	0.07

Values taken from Ref. [1].

in a given medium. The Hamaker constants are usually tabulated as A_{123} where objects 1 and 3 are interacting in the dispersing medium 2.

Example 3.8 Calculate the attraction between two 100 μm (dia) spherical particles of mica in water. The separation distances is 1 μm.

Solution: The interaction energy for small separation ($D << R$) will be applicable in this case. The Hamaker constant is 2×10^{-20} J. Hence, the interaction energy at a separation of 0.1 μm is

$$w = -\frac{2 \times 10^{-20} \times 100 \times 10^{-6}}{12 \times 1 \times 10^{-6}} \text{J} = -42kT.$$

□

It is clear from the above example that when the spheres are in close proximity, the force of attraction is very strong. Hence, the spheres cannot be dislodged by the forces of thermal fluctuations or hydrodynamic forces.

3.3 Measurement of Inter-Molecular Forces

One of the ways to measure inter-molecular forces is to measure the pressure/volume/temperature (PVT) behaviour of a real gas. Since the PVT behaviour of a real gas can be described by the van der Waals equation of state, the two parameters of the van der Waals equation, a and b, can be obtained with reasonable accuracy. Parameter a is related to the attractive interaction between the (gas) molecules and parameter b accounts for the finite molecular volumes.

As shown earlier, the van der Waals interaction is of the form

$$w(r) = -\frac{C}{r^6},$$

and the total attractive interaction energy of a molecule surrounded by other molecules is

$$\mu_{i0} = \int_{\sigma}^{\infty} w(r)\rho 4\pi r^2 dr = -\frac{4\pi C\rho}{3\sigma^3},$$

where ρ is the number of molecules per unit volume. In the above, we have assumed that only attractive interaction exists among particles, which is a reasonable approximation for hard sphere-type molecules. The repulsion is zero before contact ($r = \sigma$) and hence the above term gives the total energy of interaction correctly.

Now, writing the chemical potential of the gas molecule as a sum of energetic and entropic terms:

$$\mu_i = \mu_{i0} + kT\ln(c).$$

Note that ρ is the inverse of the volume accessible to each molecule,denoted by v_i. However, not the entire volume is accessible to the molecules (excluded volume effect) and the true number concentration considering the finite molecular volume is

$$\frac{\rho}{1-\rho v_m},$$

where v_m is the volume of a single molecule. Using the above expression, the chemical potential (per molecule basis) becomes

$$\mu_i = -\frac{4\pi C\rho}{3\sigma^3} + kT\ln(\rho/(1 - v_m\rho)).$$

We want to extract the PVT relation from this expression using the thermodynamic relation

$$\left(\frac{\partial\mu_i}{\partial P}\right)_T = v_i = \frac{1}{\rho},$$

which can be written as

$$\left(\frac{\partial\mu_i}{\partial\rho}\right)_T\left(\frac{\partial\rho}{\partial P}\right)_T = \frac{1}{\rho}$$

or

$$\left(\frac{\partial P}{\partial\rho}\right)_T = \rho\left(\frac{\partial\mu_i}{\partial\rho}\right)_T$$

or

$$\begin{aligned}
P &= \int_0^{\rho} \rho\left(\frac{\partial\mu_i}{\partial\rho}\right)_T d\rho \\
&= \int_0^{\rho} \rho\left(-\frac{4\pi C}{3\sigma^3} + \frac{kT}{\rho(1-v_m\rho)}\right) d\rho \\
&= -\frac{2\pi C}{3\sigma^3}\rho^2 - \frac{kT}{v_m}\ln(1 - v_m\rho).
\end{aligned}$$

We can expand the second term to

$$\ln(1 - v_m\rho) = -v_m\rho - \frac{(v_m\rho)^2}{2} - \cdots$$

Higher order terms can be neglected at low to moderate pressure. Hence, retaining the first- and second-order terms:

$$\ln(1 - v_m\rho) = -v_m\rho\left(1 + \frac{v_m\rho}{2}\right) = -\frac{v_m\rho}{(1 + \frac{v_m\rho}{2})^{-1}} = -\frac{v_m\rho}{(1 - \frac{v_m\rho}{2})}.$$

Using this approximation:

$$\left(P + \frac{2\pi C\rho^2}{3\sigma^3}\right) = \frac{kT}{\frac{1}{\rho} - \frac{v_m}{2}}.$$

Using $v_i = 1/\rho$:

$$\left(P + \frac{2\pi C}{3\sigma^3 v_i^2}\right)\left(v_i - \frac{v_m}{2}\right) = kT.$$

In terms of molar volume $v = Nv_i$:

$$\left(P + \frac{2\pi CN^2}{3\sigma^3 v^2}\right)\left(v - \frac{Nv_m}{2}\right) = RT.$$

The meaning of the van der Waals parameters a and b is clear: a is related to the van der Waals constant C_{vdw} and b is the excluded volume parameter. From the fitted value of a, C_{vdw} can be obtained, and the molecular diameter σ can be obtained from b.

3.4 Measurement of Forces between Surfaces

Forces between surfaces are usually measured using a spring, as shown in Figure 3.7. For the measurement, one of the surfaces remains fixed while the other is attached to a spring. If the surfaces are far enough apart, no attractive force exists and the spring remains at its equilibrium position. When the surfaces are brought closer together they attract each other and the spring deflects in response to this force. The amount of attractive force can be measured by measuring the deflection of the spring as the spring constant is known.

However, for a fixed stiffness spring, the attractive force may produce a deflection higher than that allowable for elastic deformation. In such cases plastic deformation of the spring occurs. At this point, the two objects jump into contact, as shown by the dotted line $x_1 - x_4$. The same thing happens when the object is retracted back. Hence, for the case shown, the spring cannot

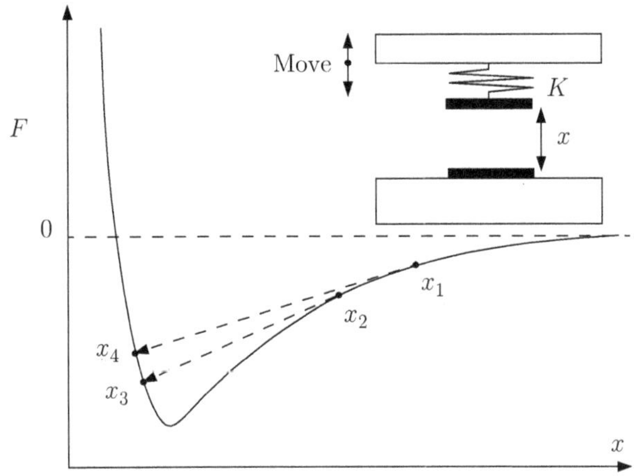

Figure 3.7 Tracing the force displacement curve using a spring.

trace the force curve between x_1 and x_4. A stiffer spring, however, will be able to trace the force curve for a wider region and will jump into contact only after the point x_2. However, it will fail to produce a measurable deflection when the forces are small.

The magnitude of the surface forces are small, at the level of milli Newtons. Hence to measure this small force a sensitive spring is required. Also, as seen from the previous paragraph, a variable stiffness spring is required to trace the entire force curve. Such variation in the spring stiffness is achieved by moving the position of the pivot of a cantilever spring.

The distance over which these forces are active is also very small, usually less than a micrometre. Hence the distance between two bodies should also be measured accurately at this length scale. In addition, the surfaces between which the force is measured must be atomically smooth. If they have roughness in the order of nanometres, the true inter-surface force cannot be measured. Mica usually offers atomically smooth surfaces without great effort and is a very popular surface for force measurements. Other test materials are usually coated on mica. Even with mica it is difficult to produce clean undulation-free flat surfaces over a large area. Hence curved surfaces are usually preferred. One of the most useful geometries is the cross cylinder.

Force between surfaces can be measured using surface force apparatus. A detailed schematic of this equipment is shown in Figure 3.8. In this apparatus precise measurement of force and distance is possible. The distance between the surfaces is controlled in three stages. First, in the submicrometre range the movement occurs via a screw similar to that used in research microscopes. The second degree of adjustment is needed in the range of 1–10 nm. This is not possible with screws. This is administered by a double cantilever or Z spring.

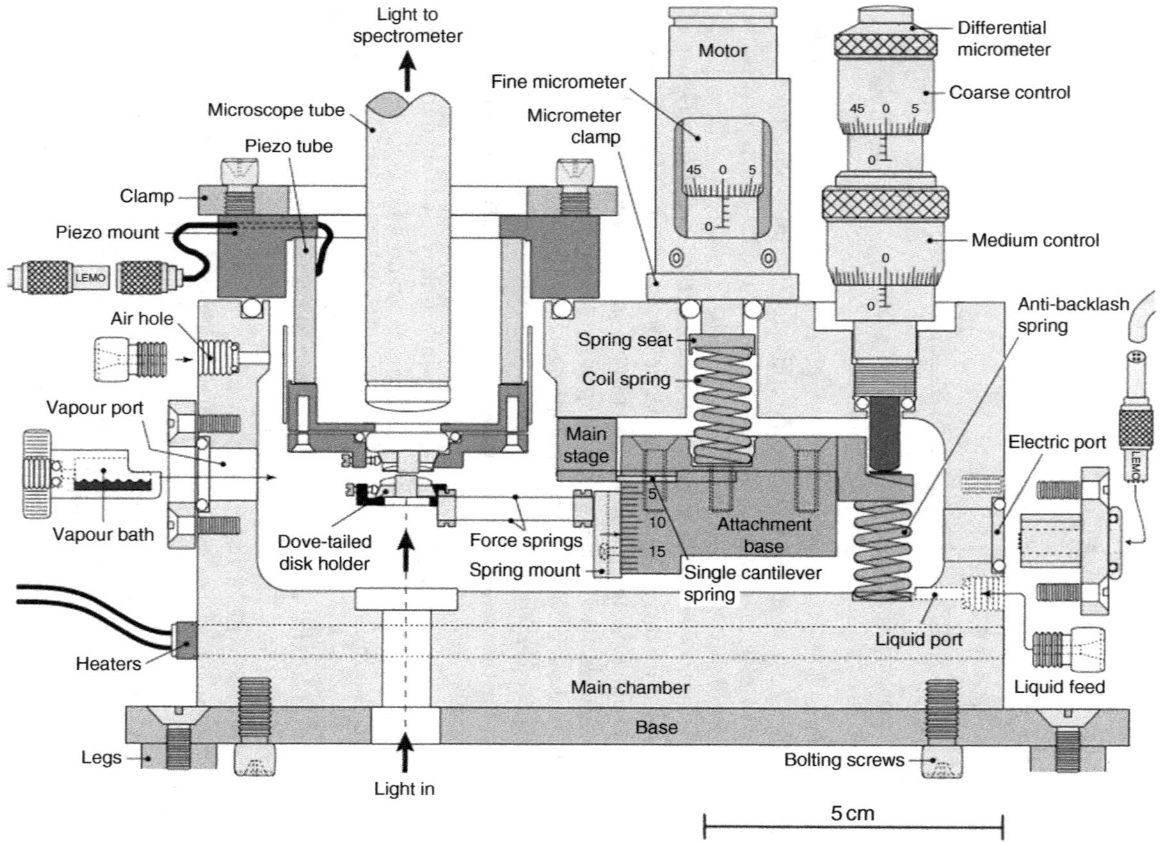

Figure 3.8 Surface force apparatus (figure taken from Ref. [2]).

Deflecting one arm of this double cantilever by 1 μm leads to a deflection of 1/1000th of this distance in the other arm. Hence, micrometre movement of a screw leads to nanometre movement of the stage using this spring. Subnanometre movement is administered by a piezoelectric device attached to the top surface.

The stages are advanced by a precise known amount. After the spring equilibrates at the new position, the distance between the surfaces is measured very accurately using interferrometry. The difference between the advancement and reduction in actual distance is the deflection of the spring. Since the spring constant is known, the force can be obtained.

3.5 Exercises

Exercises 3.1 Calculate the Hamaker constants from Hamaker's theory and compare them with those obtained using Lifshitz theory.

Exercises 3.2 Calculate the sphere–sphere interaction and show its variation with distance.

Exercises 3.3 Using a suitable expression for the attractive potential, show that the cohesive energy and melting point of methane are 9.8 kJ/mol and 182 K, respectively.

Exercises 3.4 Show that the force on a charge from an infinite surface of uniform charge density is independent of the distance from the surface.

Bibliography

1 J. N. Israelachvili. *Intermolecular and Surface Forces,* 3rd edition. Academic Press, 2011.

2 J. Israelachvili, Y. Min, M. Akbulut, A. Alig, G. Carver, W. Greene, et al. Recent advances in the surface forces apparatus (SFA) technique. *Reports on Progress in Physics, 73*(3):036601, 2010.

4

Stability

In the previous chapter we saw that small particles experience a large attractive potential. Hence, if two particles approach each other, they are attracted by a strong force and the energies of thermal fluctuation or hydrodynamic forces are not enough to separate them. Hence, the attractive potential must be balanced by a suitable repulsive potential for a well-dispersed colloid. Many natural/synthetic colloidal systems remain stable for a prolonged period because of the presence of such repulsive potentials. The repulsive potential usually originates from three major sources:

- adsorption/desorption of ions on the particle surface
- adsorbed non-ionic surfactants/polymers on the surface
- adsorbed ionic surfactant/polymer on the particle surface.

In this chapter we will quantify the repulsive force/potential between two surfaces. We will restrict ourselves to flat surfaces but the results will be valid for curved surfaces with very little modification.

Charged Interface

Charge exists in the aqueous phase because of the dissociation of electrolytes [1]. Even in the absence of electrolyte, H^+ and OH^- ions are present. Some of the existing ions may preferentially adsorb on the particle surface, which gives rise to the surface charge. In some cases the material may dissociate near the surface and release a particular ion that moves into the solution. The opposite ion remains bound with the surface and produces a surface charge.

Not all surfaces adsorb or release charges. For such surfaces, ionic surfactants may be added to the solution to produce a surface charge. The tail groups of the surfactants will be adsorbed on the particle while the counter-ions float in the solution. In all such cases, the overall system remains charge neutral. The total charge of surface bounded ions/surfactants must exactly balance the total charge of floating ions. Figure 4.1 shows such a system with a flat surface.

Engineering of Submicron Particles: Fundamental Concepts and Models, First Edition. Jayanta Chakraborty.

Companion website: http://booksupport.wiley.com

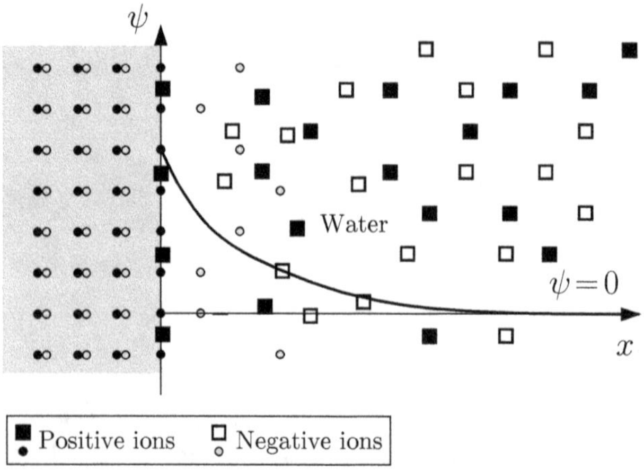

Figure 4.1 Dissociation of ions at the water–solid interface to form a charged interface.

4.1 Electrostatic Potential Near a Charged Surface

The density of the floating charges is a function of the distance from the surface. Let us denote the density of the ith ion at a distance x by $\rho_i(x)$ (number of charged entities per unit volume). The electric field due to these charges and the spatial distribution of charge ($\rho_i(x)$) are related by the Poisson equation [2]:

$$\nabla.E = \sum \frac{z_i e \rho_i}{\epsilon}. \tag{4.1}$$

Here $\epsilon = \epsilon_r \epsilon_0$, z is the valency of the ion, and e is the electronic charge. Because the field can be written as a negative gradient of a potential ($E = -d\psi/dx$), the above equation can be written in terms of the electrostatic potential for the one-dimensional case as

$$\frac{d^2\psi}{dx^2} = -\sum \frac{z_i e \rho_i}{\epsilon}.$$

However, the charge density is also related to the electrostatic potential through the Boltzmann distribution:

$$\rho_i = \rho_{i0} \exp\left(-\frac{z_i e \psi}{kT}\right).$$

Here ρ_{i0} is the concentration of the charged species where the potential is zero. For a charged surface, the potential due to the surface charge decays with the distance from the surface. Hence, in such a case ρ_{i0} is the concentration of the charged entity far away from the surface, which is approximately taken as the bulk concentration of the ion. These two equations can be combined to obtain an equation for the potential near a charged surface:

$$\frac{d^2\psi}{dx^2} = -\sum \frac{z_i e \rho_{i0}}{\epsilon} \exp\left(-\frac{z_i e\psi}{kT}\right). \tag{4.2}$$

This equation is called the Poisson–Boltzmann equation. We need to supply the boundary conditions for this equation. The potential and its derivative both become zero at a large distance from the surface. Hence, the boundary condition becomes

$$\psi\Big|_{x=\infty} = 0 \quad \text{and} \quad \frac{d\psi}{dx}\Big|_{x=\infty} = 0. \tag{4.3}$$

The boundary condition at $x = 0$ is obtained by using overall charge neutrality. Since the overall system must be charge neutral, the total number of positive and negative charges must balance exactly. Hence, if the surface charge density is σ, it must be balanced by the net amount of floating counter-ions:

$$\sigma = -\int_0^\infty \sum z_i e \rho_i(x) dx = -\epsilon \int_0^\infty -\frac{d^2\psi}{dx^2} dx = \epsilon \left[\frac{d\psi}{dx}\right]_0^\infty,$$

which gives

$$\sigma = \epsilon \left(\frac{d\psi}{dx}\Big|_{x=\infty} - \frac{d\psi}{dx}\Big|_{x=0} \right). \tag{4.4}$$

Using eqn 4.3,

$$\sigma = -\epsilon \frac{d\psi}{dx}\Big|_{x=0}. \tag{4.5}$$

The surface charge density is known from the chemistry of the surface or from independent measurement.

4.2 Solution of the Poisson–Boltzmann Equation

It can be seen that the Poisson–Boltzmann equation is a non-linear equation. However, a linearized version can be obtained for the small potential limit. If the potential energy per ion is much less than kT (which corresponds to 10 mV of surface potential and usually provides reasonable stability), the exponential term can be expanded and linearized:

$$\begin{aligned}\frac{d^2\psi}{dx^2} &\approx -\frac{e}{\epsilon} \sum_i z_i \rho_{i0} \left(1 - \frac{z_i e\psi}{kT}\right) \\ &= -\frac{e}{\epsilon} \sum_i z_i \rho_{i0} + \frac{e^2}{\epsilon} \sum_i z_i^2 \rho_{i0} \frac{\psi}{kT}.\end{aligned}$$

The concentrations ρ_{i0} refer to the concentrations far away from the surface. At this location, the effect of the surface is absent. Hence, every portion of the solution is electroneutral at this point, which yields

$$\sum_i z_i \rho_{i0} = 0.$$

Hence, the above equation becomes

$$\frac{d^2\psi}{dx^2} = \frac{e^2}{\epsilon} \sum_i z_i^2 \rho_{i0} \frac{e\psi}{kT} = \left[\sum_i \rho_{i0} \frac{z_i^2 e^2}{kT\epsilon} \right] \psi.$$

The term in square brackets (denoted by κ^2) has a dimension L^{-2} and hence provides the natural length scale for the system. Hence, the above equation becomes

$$\frac{d^2\psi}{dx^2} = \kappa^2 \psi.$$

This equation can be readily solved and the solution is

$$\psi = \psi_0 \exp(-\kappa x). \tag{4.6}$$

Here the surface potential is given by ψ_0. The significance of κ is clear now: it is similar to the time constant for first-order kinetics. The potential drops by a factor e at a distance $1/\kappa$ from the surface. This length is known as the Debye length and all distances near a charged interface can be scaled with this length scale. The surface potential ψ_0 can be related to the surface charge density

$$\left. \frac{d\psi}{dx} \right|_{x=0} = - \left. \kappa\psi_0 \exp(-\kappa x) \right|_{x=0} = -\kappa\psi_0$$

using eqn 4.5

$$\psi_0 = \frac{\sigma}{\kappa\epsilon}.$$

If the surface potential is not smaller than kT, the exponential series cannot be truncated. The non-linear equation can still be solved for the restricted case of a symmetric electrolyte with valency z and the solution is expressed as

$$\gamma = \gamma_0 \exp(-\kappa x), \tag{4.7a}$$

where

$$\gamma = \frac{\exp\left(\frac{ze\psi}{2kT}\right) - 1}{\exp\left(\frac{ze\psi}{2kT}\right) + 1} \tag{4.7b}$$

and

$$\gamma_0 = \frac{\exp\left(\frac{ze\psi_0}{2kT}\right) - 1}{\exp\left(\frac{ze\psi_0}{2kT}\right) + 1} \tag{4.7c}$$

are dimensionless potentials at any location and surface, respectively.

Table 4.1 Solutions of the Poisson–Boltzmann equation and their domain of applications

Name	Applies when	Solution for ψ
Debye–Huckel	Small surface potential ($ze\psi/kt << 1$); applies in the entire domain	Equation 4.6
Guoy–Chapman	Symmetric electrolyte. Any surface potential; entire domain	Equation 4.7
Guoy–Chapman ($x >> \kappa$)	Any surface potential; large distance from the surface	Equation 4.8

Irrespective of the magnitude of the surface potential, the potential drops to a small value (with respect to kT) at large distance from the surface. Hence, for $x >> \kappa$ the exponentials in γ may be expanded in Taylor's series and higher terms can be neglected:

$$\frac{\exp\left(\frac{ze\psi}{2kT}\right)-1}{\exp\left(\frac{ze\psi}{2kT}\right)+1} = \frac{\frac{ze\psi}{2kT}}{\frac{ze\psi}{2kT}+2}.$$

The term $\frac{ze\psi}{2kT}$ is much smaller than 1. Hence, the spatial variation for potential at a large distance from the surface can be written as

$$\psi = \frac{4kT\gamma_0}{ze}\exp(-\kappa x) \quad (x >> \kappa). \tag{4.8}$$

Various solutions of Poisson–Boltzmann equations are summarized in Table 4.1.

Example 4.1 Surface potential is a directly measurable quantity. For a surface, it is measured to be 10 mV at 300 K. Can this case be treated under the Debye–Huckel theory?

Solution: For the Debye–Huckel theory to apply, the surface potential must be less than kT or $\frac{e\psi_0}{kT} << 1$. In this case,

$$\frac{e\psi_0}{kT} = \frac{1.6\times10^{-19}\times10\times10^{-3}}{1.38\times10^{-23}\times300} = 0.386.$$

Hence, the Debye–Huckel theory can be used. □

Example 4.2 What is the Debye length of a surface which is dipped in 10 mM NaCl solution?

Solution: In this case, $z_1 = +1$ and $z_2 = -1$, $n_{1\infty} = n_{2\infty} = n_\infty$. Hence,

$$\kappa^2 = \frac{e^2}{\epsilon kT}\sum_{i=1,2} z_i^2 n_{i\infty} = \frac{2e^2 n_\infty}{\epsilon kT}$$
$$= \frac{2\times(1.6\times10^{-19})^2\times10\times10^{-3}\times10^3\times6.023\times10^{23}}{80\times8.85\times10^{-12}\times1.38\times10^{-23}\times300} = 6.57\times10^{16},$$

which gives $\kappa = 2.5\times10^8$. Hence, the Debye length is $\kappa^{-1} = 3.9\times10^{-9}$ m or ≈ 4 nm. □

4.3 Repulsive Force between Two Surfaces

In the previous section we obtained the electric potential near an isolated surface. Because there exists a gradient in the potential, each of the charged particles in the potential field will experience a force. This situation is similar to the gravitational force experienced by fluid molecules on the surface of the earth. Because the fluid remains static in a gravitational (as well as an electric) field, the force due to the gravitational (or electric) field must be balanced by a pressure force. In establishing the parallel between the gravitational and electric fields, we implicitly assume that the collection of charge behaves like a continuum. Usually electrolytes are present in large amounts and this is a reasonable assumption.

The electrostatic force per unit volume in the x direction is given by the charge density multiplied by the electric field:

$$\left(\sum \rho_i z_i e\right)\times E = -\left(\sum \rho_i z_i e\right)\frac{d\psi}{dx}.$$

This force must be balanced by the pressure force per unit volume. This can be obtained by considering an elementary volume, as shown in Figure 4.2. The net x direction pressure force acting on the fluid of elementary volume $dv = dxdydz$ is

$$p(x)dydz - \left(p(x) + \frac{dp}{dx}dx\right)dydz = -\frac{dp}{dx}dxdydz = -\frac{dp}{dx}dv.$$

Hence, the net x directional pressure force per unit volume is simply $-dp/dx$. Since these two forces must be equal and opposite,

$$\frac{dp}{dx} + \left(\sum \rho_i z_i e\right)\frac{d\psi}{dx} = 0.$$

Using Poisson's equation (eqn 4.1), the above equation becomes

$$\frac{dp}{dx} - \epsilon\frac{d^2\psi}{dx^2}\frac{d\psi}{dx} = 0$$

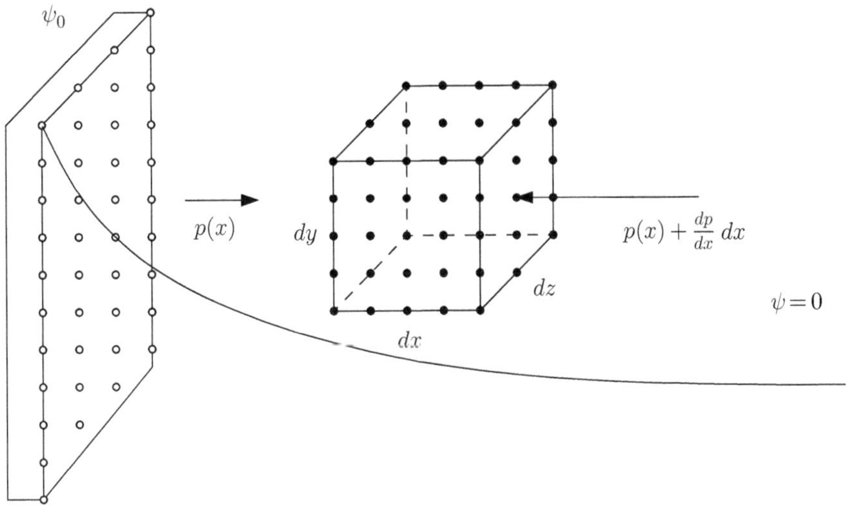

Figure 4.2 An elementary volume near a charged surface.

or

$$\frac{dp}{dx} - \frac{\epsilon}{2}\frac{d}{dx}\left(\frac{d\psi}{dx}\right)^2 = 0,$$

which gives

$$p - \frac{\epsilon}{2}\left(\frac{d\psi}{dx}\right)^2 = \text{constant}. \tag{4.9}$$

A boundary condition is needed to evaluate the constant. At a large distance from the surface, $d\psi/dx = 0$. At this location the pressure is the overall system pressure and is denoted as p_0. Hence,

$$p = p_0 + \frac{\epsilon}{2}\left(\frac{d\psi}{dx}\right)^2.$$

The above equation shows that pressure gradually increases towards the surface.

Now consider two parallel plates of infinite extent. Both sides of both the plates have identical surface conditions and the electrolyte is identical on all sides. Let us consider a situation where the electrostatic potential of the two plates just started to overlap. This situation is shown in Figure 4.3. If we consider the right side of plate 2 or the left side of plate 1, the situation is identical to that discussed above and the surface pressures can be written as

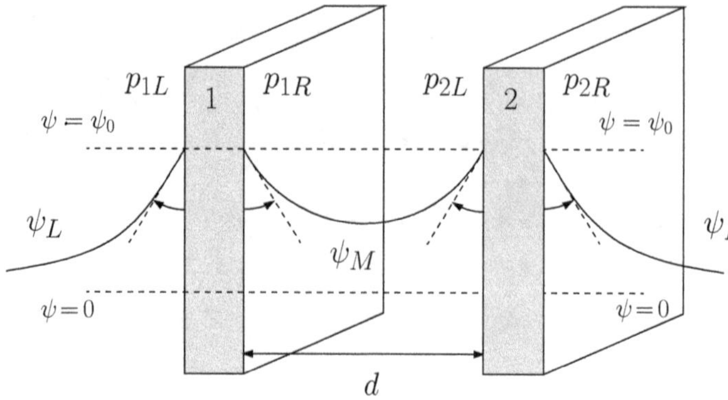

Figure 4.3 Surface pressure.

$$p_{2R} = p_0 + \frac{\epsilon}{2}\left(\left.\frac{d\psi}{dx}\right|_{2R}\right)^2$$

$$p_{1L} = p_0 + \frac{\epsilon}{2}\left(\left.\frac{d\psi}{dx}\right|_{1L}\right)^2.$$

However, in between the plates the situation is slightly different. The quantity $d\psi/dx$ drops to zero at a distance $d/2$ from the surface instead of at a large distance. Hence, the constant of integration will be different in this case. It can readily be seen that it will be the pressure at the mid-plane that is denoted as $p_{d/2}$. Hence, the pressure at the right side of plate 1 and left side of plate 2 can be written as

$$p_{1R} = p_{d/2} + \frac{\epsilon}{2}\left(\left.\frac{d\psi}{dx}\right|_{1R}\right)^2$$

$$p_{2L} = p_{d/2} + \frac{\epsilon}{2}\left(\left.\frac{d\psi}{dx}\right|_{2L}\right)^2.$$

As shown earlier (see eqn 4.5), the quantity $d\psi/dx|_{\text{Surface}}$ is dependent only on the surface charge density. Hence, they are the same on both the left and right sides of both the surfaces. Hence, the net outward pressure on plate 1 is given by

$$p_{1L} - p_{1R} = p_{d/2} - p_0.$$

We need to express $p_{d/2}$ in terms of known quantities. Equation 4.9 cannot be used for this purpose because we used this equation as a defining equation for $p_{d/2}$. However, in all cases the original differential force balance equation remains valid and can be used. This equation can be written as

$$dp + \left(\sum \rho_i z_i e\right) d\psi = 0.$$

Let us first eliminate ρ_i using the Boltzmann equation and also restrict ourselves to symmetric 1:1 electrolytes:

$$\sum \rho_i z_i e = -ze\rho_0 \exp(ze\psi/kT) + ze\rho_0 \exp(-ze\psi/kT) = -2ze\rho_0 \sinh(ze\psi/kT)$$

Using the above relation the force balance becomes

$$dp = 2ze\rho_0 \sinh(ze\psi/kT)d\psi.$$

Now imagine a process where we bring the two plates shown in Figure 4.3 from infinite separation to the separation distance d. The potential at the mid-plane changes from $\psi = 0$ to $\psi_{d/2}$ during this process and the mid-plane pressure also changes from p_0 to $p_{d/2}$. Integrating between these limits,

$$p_{d/2} - p_0 = 2kT\rho_0 \left[\cosh\left(\frac{ze\psi_{d/2}}{kT}\right) - 1\right].$$

For small overlap, i.e. at the beginning of the approach, the total potential is much smaller than kT. In this case, the cosh term can be expanded in series and the higher-order terms may be neglected:

$$F_R = p_{d/2} - p_0 = 2kT\rho_0 \left[1 + \left(\frac{ze\psi_{d/2}}{kT}\right)^2 \Big/ 2! + \left(\frac{ze\psi_{d/2}}{kT}\right)^4 \Big/ 4! + \cdots - 1\right]$$

$$= 2kT\rho_0 \left[\left(\frac{ze\psi_{d/2}}{kT}\right)^2 \Big/ 2!\right] = \rho_0 \frac{z^2e^2\psi_{d/2}^2}{kT}.$$

For small potential, the potential at the mid-plane can be taken as the algebraic sum of the potentials at the mid-plane:

$$\psi_{d/2} = \psi_1 + \psi_2 \approx 2\psi_1 = 2\frac{4kT\gamma_0}{ze} \exp(-\kappa d/2).$$

In the above, we have used eqn 4.7. Substituting this in the above equation:

$$F_R = 64\rho_0 kT\gamma_0^2 \exp(-\kappa d).$$

The distance-dependant potential can be obtained by integrating the force:

$$\Phi(d) = -\int_{\infty}^{d} F_R(y)dy = \frac{64\rho_0 kT\gamma_0^2}{\kappa} \exp(-\kappa d).$$

This is the repulsive potential per unit area between the two flat surfaces when maintained at a distance d. It can be shown (see Exercise 4.2) that for spheres of equal radius the potential becomes

$$\Phi_R(d) = \frac{64\pi R\rho_0 kT\gamma_0^2}{\kappa^2} \exp(-\kappa d). \quad (4.10)$$

Example 4.3 Compute the repulsive potential between two silver particles of 50 nm at various distances in 16 mM AgBr solution. The surface potential is measured to be 25 mV.

Solution: The electrolyte is a 1:1 symmetric electrolyte. For water at 300 K the factor $e^2/(\epsilon kT)$ is 17×10^{-9}. The number concentration of electrolyte at a large distance from the surface is $\rho_0 = 16 \times 10^{-3} \times 6.023 \times 10^{23} \times 10^3 = 0.963 \times 10^{25}$. Hence, κ^2 for this case is

$$\kappa^2 = 17 \times 10^{-9} \times 0.963 \times 10^{25} = 1.637 \times 10^{17},$$

which gives $\kappa = 4 \times 10^8\ \text{m}^{-1}$ or $\kappa^{-1} = 2.5$ nm. For a surface potential of 25 mV,

$$\frac{ze\psi_0}{2kT} = 0.483,$$

which gives $\gamma_0 = 0.237$. Hence, using eqn 4.10,

$$\begin{aligned}\frac{\Phi(d)}{kT} &= \frac{64\pi R\rho_0\gamma_0^2}{\kappa^2}\exp(-\kappa d)\\ &= \frac{64\pi \times 50 \times 10^{-9} \times 0.963 \times 10^{25} \times 0.237^2}{1.637 \times 10^{17}}\exp(-\kappa d)\\ &= 33\exp(-\kappa d).\end{aligned}$$

Hence, the following table can be prepared:

d/κ^{-1}	ϕ/kT
1	12
2	4.46
3	1.64
4	0.60

It can be seen that the repulsive potential is tens of kT at a separation of 2.5 nm, but the attractive potential at this distance will be higher and the particle will aggregate. □

Example 4.4 Consider an infinitely long flat charged surface of uniform charge density σ. It is in contact with water where only counter-ions exist in the system. If the counter-ions are arranged in a single layer at a distance κ^{-1} from the surface, as shown in Figure 4.4, plot the potential and the electric field as a function of distance from the surface. Explain your plot.

Solution: It can be shown using Coulomb's law that the electric field due to an infinitely long surface of uniform charge density is independent of distance from the surface and is given by $\sigma/(2\epsilon)$ (see Exercise 3.4). Here we have two such plains. The field is additive in the space between the two layers of uniform charge. According to the coordinate system shown in the figure, the resulting field acts in the positive x direction. Hence, the field is σ/ϵ in between the two layers and zero outside the two layers.

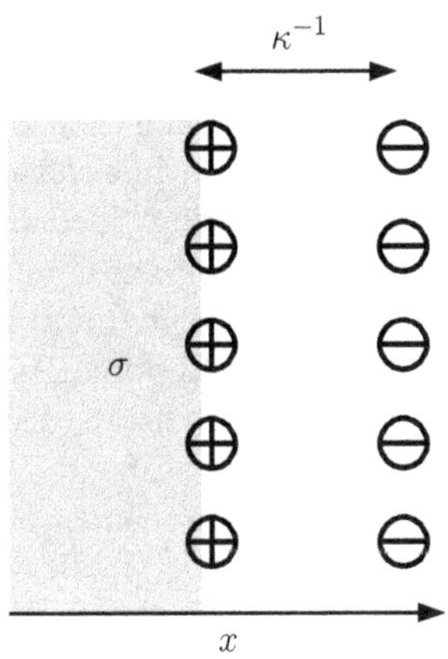

Figure 4.4 Counter-ions arranged in a single layer at a distance κ^{-1} from the surface.

The electric field is the negative gradient of the potential. Hence, the equation for the potential for this case can be written as

$$-\frac{d\psi}{dx} = \frac{\sigma}{\epsilon}.$$

This shows a linear plot with a negative slope of magnitude σ/ϵ. □

4.4 Steric Stabilization

The charged stabilization discussed above works well for many aqueous systems. Charged stabilized colloids may be stable for years. However, the main drawback for charged stabilization is that it is dependent on ionic concentrations. If the salt concentration is higher, the colloid becomes unstable. It also works only for hydrocolloids. To stabilize colloidal systems where the charge stabilization is either inadequate or not possible, steric stabilization is used [3, 4]. For steric stabilization, proteins, surfactants or polymers are adsorbed on the surface instead of ions, and steric force is used to create the activation barrier.

Proteins, surfactants or polymers are long-chain molecules part of which can be pinned to the particle surface by chemical or physical bonding. The remaining chain dangles in the fluid. The part that attached to the particle is called the anchoring group and other part is the stabilization moiety. When two particles with attached stabilizer molecules are close together, the dangling portions of the molecules have restricted movement. This is entropically unfavorable and provides repulsion.

The concentration of polymer used in steric stabilization is extremely important. If the polymer concentration is very low or the polymer contains multiple anchoring groups, one part of the same polymer molecule may attach on one particle while the other part attaches to another particle. This situation is called bridging flocculation. At higher concentrations the probability of multiple attachment is much less because the number of polymer molecules per particle is higher and bridging is very rare. At this point, particle surfaces are covered with polymers, which leads to a stable colloid.

When the polymer concentration increases above this limit, there is adsorbed polymer on the surface as well as a high concentration of polymer molecules in the solution. Upon close approach of two particles, the space between the two particles becomes smaller than the dimension of the polymer. Hence, this space becomes inaccessible by the polymers. This again becomes entropically unfavorable. The system wants to eliminate this space by closing in and aggregation occurs. This is called depletion flocculation.

4.5 Kinetics of Stability

In the previous sections we have seen that there exist both attractive and repulsive potentials between particles. When two particles approach each other, the net potential between them governs the nature of the aggregation process. Various possibilities that can arise depending on the shape of the net potential curve are shown in Figure 4.5. Usually, the net potential is obtained by summing the van der Waals attraction and double layer repulsion. This was first explored by Boris Derjaguin, Lev Landau, Evert Verwey, and Theo Overbeek, and is known as DLVO theory [5].

It can be seen that in absence of a repulsive potential, particles jump into a deep primary minimum (usually hundreds of kT deep) and it is virtually impossible to dislodge such aggregates. On the other hand, if the repulsive barrier is very strong, a large activation barrier prevents such aggregation and particles remain well separated.

In many cases, however, the repulsive barrier is only moderate. In such cases (see Exercise 4.1) a small secondary minimum is seen before the activation barrier. Such a minimum is usually only a few kT deep and hydrodynamic forces are enough to dislodge aggregates formed due to this energy minimum. Such reversible aggregates are called flocs.

Although the particles are shielded from the primary minimum by an activation barrier a fraction of the particles usually cross this barrier due to diffusive movement and form an aggregate. In this section we will examine the diffusive movement of particles in the presence of a net potential. The aim is to predict the duration for which a particulate system remains in its dispersed state in the presence of a given potential.

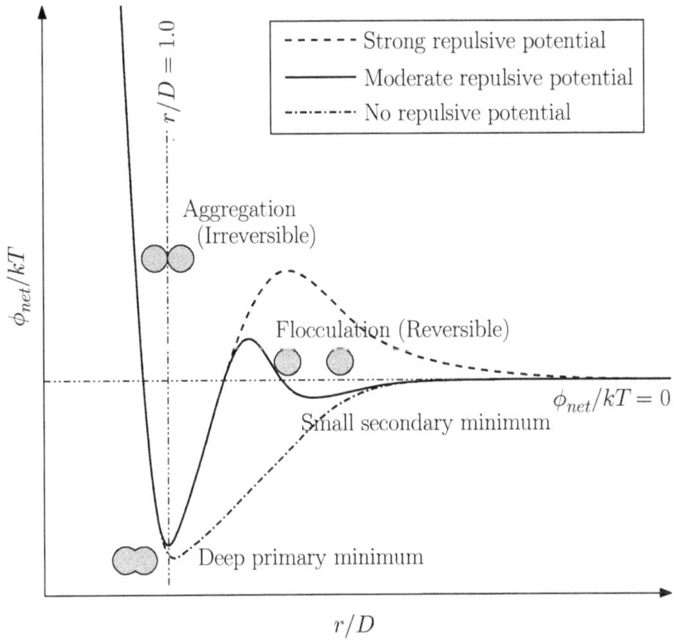

Figure 4.5 Various possible landscapes for the net potential.

4.5.1 Diffusion of Colloidal Particles

Engineers are familiar with the phenomena of molecular diffusion. Diffusion of particles is very similar to diffusion of molecules but with a very prominent difference: every molecule of a chemical species are of same size but particles are usually of different sizes. Particle size has a strong effect on diffusivity, as discussed below.

The diffusivity of particles is best analyzed by imagining the dispersed particles of a given size as a species in the solution and considering the chemical force acting on the particles. A particular species diffuses in a solution because of the spatial gradient in its chemical potential. Let us denote the chemical potential (per particle) for particle type i (defined by size, chemical composition etc.) as μ_i, which is given by

$$\mu_i \approx \mu_i^0 + kT \ln c_i.$$

Since a force can be defined as the negative gradient of the potential, the chemical force acting on a particle can be written as

$$F_c = -\frac{d\mu_i}{dx} = -\frac{kT}{c_i}\frac{dc_i}{dx}.$$

If we assume that the particle achieves a constant velocity during its flight, this force will be balanced by the viscous drag. The drag force on a spherical particle of radius R_s is given by $6\pi R_s u\mu$ and therefore

$$6\pi R_s u\mu = -\frac{kT}{c_i}\frac{dc_i}{dx}. \tag{4.11}$$

The diffusivity for particles can be defined using Fick's law:

$$J = uc_i = -D\frac{dc_i}{dx}. \tag{4.12}$$

Using eqn 4.12 to eliminate $(1/c_i)dc_i/dx$ from eqn 4.11 we obtain

$$D = \frac{kT}{6\pi\mu R_s}. \tag{4.13}$$

Hence, the diffusive movement of colloidal particles is inversely proportional to their size.

4.5.2 Particle Aggregation in the Absence of Potential

The diffusive movement of particles brings particles into close proximity. If no potential barrier is present, particles will latch onto the primary minimum irreversibly. Hence, each time two particles collide with each other, one particle is lost from the system. The rate at which particles are lost from the system is therefore given by the rate of collision.

Consider a particle with radius R_2, as shown in Figure 4.6. We are interested to know how many particles of type 1 (with radius R_1) will collide with this particle within a small time window. We will assume the central particle to be stationary and purely diffusive movement of particle type 1. The analysis of this situation is identical to that shown in section 2.1.1 and the pseudo steady state concentration distribution of particles of type 1 can be written as

$$c_1(r) = -\frac{c_{1\infty}(R_1 + R_2)}{r} + c_{1\infty}.$$

In the above we also assume that the particle concentration at the surface is zero. This means that the colliding particles merge instantaneously. From this expression, the flux at the contact plane (the inner dotted circle) can be calculated as

$$J_c = -D_1\frac{dc_1}{dr}\bigg|_{r=R_1+R_2} = -D_1\frac{c_{1\infty}}{R_1 + R_2}$$

and the total number of collisions per unit time at the contact plain becomes

$$J_c \times 4\pi(R_1 + R_2)^2 = -4\pi D_1 c_{1\infty}(R_1 + R_2).$$

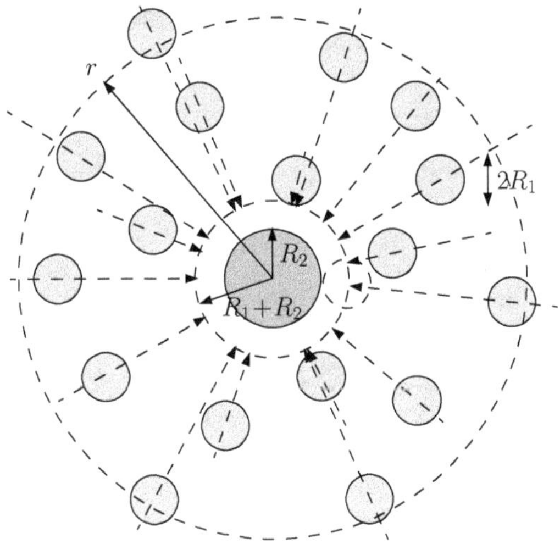

Figure 4.6 Diffusion of particles with radius R_1 towards a particle with radius R_2.

The negative sign signifies that the flux is in the negative r direction. This provides the number of collisions for a single particle of size R_2. However, there will be large number of such particles and the total number of collisions for all particles of type 2 will be

$$4\pi D_1 c_{1\infty} c_{2\infty}(R_1 + R_2).$$

Up to this point we have considered the central particle to be stationary. If the central particle moves, we can view this situation as diffusion of particle type 2 towards particle type 1. The total rate for this can be obtained in an identical manner and the total number of collisions in this case will be

$$4\pi D_2 c_{1\infty} c_{2\infty}(R_1 + R_2).$$

When both the particles are diffusing, the net rate of collision per unit volume can be approximated as the sum of the above quantities and is given by

$$\frac{dc_{1\infty}}{dt} = -4\pi(D_1 + D_2)c_{1\infty} c_{2\infty}(R_1 + R_2).$$

For particles of identical size ($R_1 = R_2 = R_s$; $\quad D_1 = D_2 = D$; $\quad c_{1\infty} = c_{2\infty} = c$),

$$\frac{dc}{dt} = -8\pi Dc^2(2R_s) \times \frac{1}{2}.$$

The factor 1/2 eliminates double counting emanated from counting the same particle once as type 1 and again as type 2. Substituting the expression for diffusivity from eqn 4.13,

$$\frac{dc}{dt} = -\frac{8kT}{3\mu}c^2 \times \frac{1}{2}.$$

This equation is used to estimate the half-life of the unprotected colloid and is usually written in the form

$$\frac{dc}{dt} = -k_r c^2,$$

with the initial condition that $c = c_0$ at $t = 0$. The solution becomes

$$\frac{1}{c} - \frac{1}{c_0} = k_r t.$$

Defining $t_{1/2}$ as the time when $c = c_0/2$,

$$t_{1/2} = \frac{1}{c_0 k_r}.$$

Example 4.5 Estimate the half-life of a colloid which has initial particle concentration 10^{10}/cc at a temperature of 50 °C. The dispersing medium is water with viscosity 8×10^{-4} Pas.

Solution:

$$k_r = \frac{4 \times 1.38 \times 10^{-23} \times 323}{3 \times 8 \times 10^{-4}} = 7.5 \times 10^{-18}$$

The particle concentration is $10^{10}/\text{cc} = 10^{16}/\text{m}^3$.

$\therefore t_{1/2} = 13$ s. If the colloid is cooled to 5 °C and diluted 10 times, the half-life would be about 160 s. No major change occurs despite the large change in the dilution and temperature. □

4.5.3 Particle Aggregation in the Presence of a Net Potential

If a repulsive potential is present, the half-life of a colloid can be improved substantially. To calculate the rate of collision in the presence of a net potential (Φ), we need to add the flux due to the potential to the pure diffusive flux. The procedure for obtaining the flux generated by a potential is very similar to that used for a chemical potential driven flux. First, we obtain the force acting on a particle due to Φ as

$$-\frac{d\Phi}{dr}.$$

The particle velocity due to this force can be calculated by equating this force with the drag force:

$$-\frac{d\Phi}{dr} = 6\pi \mu u R_s.$$

The corresponding flux can be calculated as $(u \times c)$

$$-\frac{d\Phi}{dr}\frac{c}{6\pi \mu R_s}.$$

The particle diffusivity, $D = kT/(6\pi\mu R_s)$, can be substituted in the above expression to obtain the potential driven flux as

$$\frac{-Dc}{kT}\frac{d\Phi}{dr}.$$

Hence, the total flux at any r is given by

$$J_T(r) = -D\frac{dc}{dr} - \frac{Dc}{kT}\frac{d\Phi}{dr}.$$

Now, at steady state, although the flux varies with r, the total number of particles crossing a spherical surface at any r will be constant. This number equals the number of collision events per unit volume per particle type 2 and is given by

$$J_T A = -4\pi r^2 D\left[\frac{dc}{dr} + \frac{c}{kT}\frac{d\Phi}{dr}\right].$$

where A is the surface of a sphere at r. The functionality has been omitted as a reminder that the quantity $J_T A$ is independent of r. If we consider movement of both the particles, the diffusivity must be doubled as shown before:

$$J_T A = -8\pi r^2 D\left[\frac{dc}{dr} + \frac{c}{kT}\frac{d\Phi}{dr}\right].$$

Rearranging,

$$\frac{dc}{dr} + \frac{c}{kT}\frac{d\Phi}{dr} = -\frac{J_T A}{8\pi r^2 D}.$$

Since $J_T A$ in the above equation is not a function of r, the equation becomes a linear equation in c whose general solution is

$$c \times \exp\left(\frac{\Phi}{kT}\right) = -\int \frac{J_T A}{8\pi r^2 D}\exp\left(\frac{\Phi}{kT}\right) dr + C$$

Although one boundary condition is needed for this first-order equation, traditionally two conditions are used to evaluate C and solve for $J_T A$ simultaneously. The two conditions are

$$@r = \infty, \quad c = c_\infty, \quad \Phi = 0$$

and

$$@r = 2R_s, \quad c = 0,$$

which yields

$$c_\infty = \left[-\int \frac{J_T A}{8\pi r^2 D}\exp\left(\frac{\Phi}{kT}\right) dr\right]_\infty + C$$

and

$$0 = \left[-\int \frac{J_T A}{8\pi r^2 D}\exp\left(\frac{\Phi}{kT}\right) dr\right]_{2R_s} + C.$$

Combining:

$$c_\infty = -\int_{2R_s}^{\infty} \frac{J_T A}{8\pi r^2 D} \exp\left(\frac{\Phi}{kT}\right) dr$$

or

$$\frac{dc_\infty}{dt} = J_T A = -\frac{8\pi D c_\infty}{\int_{2R_s}^{\infty} \frac{1}{r^2} \exp\left(\frac{\Phi}{kT}\right) dr}.$$

The above expression is for a single particle. Considering all particles and removing double counting:

$$\left.\frac{dc_\infty}{dt}\right|_{Slow} = -\frac{1}{2}\frac{8\pi D c_\infty^2}{\int_{2R_s}^{\infty} \frac{1}{r^2} \exp\left(\frac{\Phi}{kT}\right) dr}.$$

The rate of coagulation without any repulsive potential is

$$\left.\frac{dc_\infty}{dt}\right|_{Fast} = -\frac{1}{2}(8kT/3\mu)c_\infty^2 = -\frac{1}{2}(2R_s)8\pi D c_\infty^2.$$

The ratio of coagulation without any potential (fast) and with potential (presumably repulsive, hence slow) is called the stability ratio. According to the above expressions, it is given by

$$W = 2R_s \int_{2R_s}^{\infty} \frac{1}{r^2} \exp\left(\frac{\Phi}{kT}\right) dr.$$

This is the practically important property of a colloid. It can be seen that it may be evaluated from known data. To estimate the stability ratio for Exercise 4.1, the following line may be run in Matlab after executing the code in Appendix C1: `W=2*R*integral(@(rcc)pot(rcc),2*R,Inf)`.

4.6 Measurement of Surface Potential

If an electric field is applied, ions move. The force on an ion of charge q in an electric field E is

$$F = q.E.$$

The movement of the ion reaches steady state very quickly and at this steady state the electrostatic force equals the drag force:

$$q.E = 6\pi\mu u R_s.$$

This gives the steady velocity of the ion as

$$u = \frac{q.E}{6\pi\mu R_s}.$$

The mobility of the ion is defined as u/E and hence the mobility is given by

$$u/E = \frac{ze}{6\pi\mu R_s}.$$

If we could visualize and measure the mobility of the ion, we could determine its charge. A similar strategy is used for measuring the surface charge/potential of a particle. For particles, the situation is more complicated. Instead of a single charged species, a particle contains many charges on its surface and counter-ions distributed in the solution. When a particle is static, the counter-ions achieve a set distribution, but when the particle moves, the movement of fluid with respect to the particle will sweep away some or most of the counter-ions.

As we have seen before, the ionic distribution reaches its bulk value within a few nanometres. Hence, the sweeping of ions depends on the velocity of fluid inside the electric double layer (EDL). In other words, it depends on the relative thickness of the hydrodynamic boundary layer (HBL) to the EDL. If the HBL is much thinner than the EDL, most of the EDL experiences the free stream velocity and the counter-ions will be swept away. On the other hand, if the HBL is much thicker than the EDL, a significant portion of the EDL experiences a small velocity and hence most of the counter-ions remain distributed around the particle.

The order of magnitude of the HBL around a particle is that of the particle radius. Hence, we shall classify the situation into two categories: $\kappa^{-1} >> R_s$ and $\kappa^{-1} << R_s$. Let us first consider the case where $\kappa^{-1} >> R_s$ or $\kappa R_s << 1$. We will consider a single charged particle floating in the solution. When such a particle is observed under microscope, it is found that it moves at a steady velocity. This velocity is measurable and we will relate it to its surface potential.

4.6.1 Surface Potential When $R_s << \kappa^{-1}$

As discussed above, most of the EDL experiences free stream velocity in this case and hence most of the counter-ions are swept away from the vicinity of the particles. The particle may be imagined as a sphere carrying a uniform distribution of charge on its surface with no counter-ion in its vicinity. This situation is shown in Figure 4.7. To analyze the situation, we will need the Poisson–Boltzmann equation for spherical geometry given by

$$\frac{1}{r^2}\frac{d}{dr}\left(r^2\frac{d\psi}{dr}\right) = \frac{e^2}{\epsilon kT}\left(\sum_i z_i^2 n_{i0}\right)\psi = \kappa^2\psi.$$

With the substitution $r\psi = x$ this equation reduces to the planer form:

$$\frac{d^2x}{dr^2} = \kappa^2 x.$$

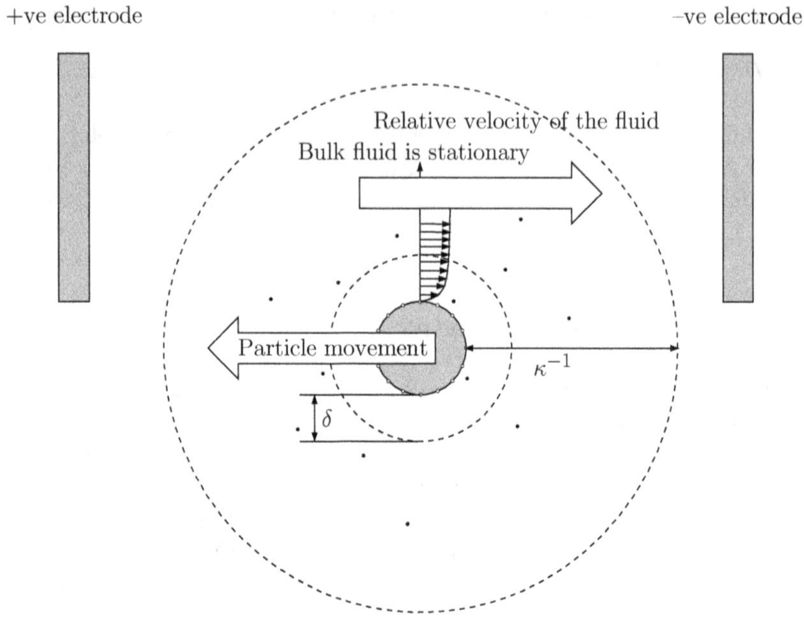

Figure 4.7 ζ potential: the case when $R_s << \kappa^{-1}$. The surface charge is negative. The velocity of fluid shown is with respect to an observer moving with the same velocity as the particle.

Solving and substituting back:

$$\psi = \frac{A\exp(-\kappa r)}{r} + \frac{B\exp(\kappa r)}{r}. \tag{4.14}$$

The constant B is evaluated readily as $r \to \infty, \psi \to 0 \Rightarrow B = 0$.

The other boundary condition is specified at $r = R_s$. Gauss law says that a spherically symmetric surface of uniform charge density may be considered as a net point charge located at the centre of a sphere. This situation applies near the surface where no counter-ion or electrolyte is present. Let us denote the net charge of the sphere as q. Hence, the electric field just outside the surface will be

$$E|_{R_s} = \frac{q}{4\pi\epsilon R_s^2}.$$

The corresponding potential is given by:

$$\left.\frac{d\psi}{dr}\right|_{R_s} = -\frac{q}{4\pi\epsilon R_s^2}.$$

Using this boundary condition and remembering that $\kappa R_s << 1$, we get

$$A = q/4\pi\epsilon.$$

Finally,

$$\psi = \frac{q}{4\pi\epsilon r}\exp(-\kappa r).$$

If the potential at surface is denoted by ζ,

$$\zeta = \frac{q}{4\pi\epsilon R_s}\exp(-\kappa R_s).$$

Also κR is very small according to our previous assumption. Therefore, the exponential term is close to unity. This gives

$$\zeta = \frac{q}{4\pi\epsilon R_s}.$$

Note that κr is not small at all values of r. This relates the surface potential to the total amount of surface charge for a spherical particle. Next we will relate the charge on the particle to its velocity. Since the particle is effectively an macro-ion, the development in the previous section can be used and the mobility of the particle becomes

$$\frac{u}{E} = \frac{q}{6\pi\mu R_s}.$$

Combining:

$$\zeta = \frac{3\mu(u/E)}{2\epsilon}.$$

Hence, the surface potential (ζ) can be obtained by measuring the velocity of the particle under a known electric field. It should be noted that the surface potential measured by the above method is somewhat lower than the true surface potential [5, 6] because of the finite ion size at surface. Hence, this potential is called the ζ (zeta) potential and is considered as an approximate measure of the true surface potential.

4.6.2 Surface Potential When $R_s >> \kappa^{-1}$

In the previous case most of the counter-ions experience the free stream velocity and are swept away from the particles. However, if the EDL is very thin compared to the particle size (see Figure 4.8), the opposite occurs: most of the ions remain attached to the particle. The fluid velocity is small near the surface and hence the viscous force is not enough to sweep away most of the counter-ions. Near the surface, the viscous force is rather balanced by the *external electric field* to maintain a stationary condition around the particle.

The problem can be described in a Lagrangian as well as in a Eulerian frame of reference. Here we shall adopt the Eulerian approach. The velocity of the fluid to a stationary observer is shown in Figure 4.9. For the charges in the thin section to travel at constant velocity, the net force on them must be zero. The

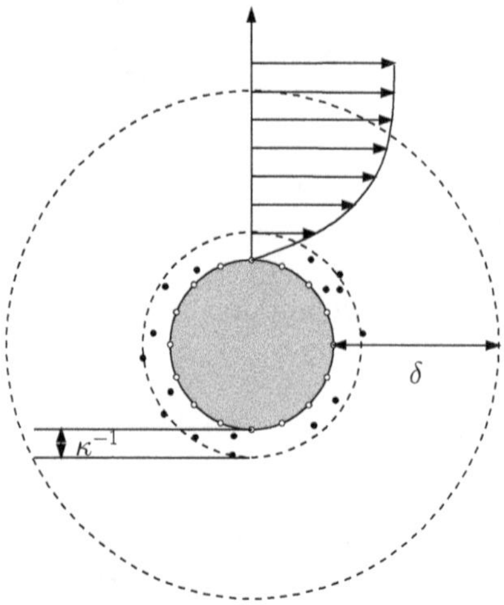

Figure 4.8 ζ potential: the case when $R >> \kappa^{-1}$. The surface charge is negative. The velocity of fluid shown is with respect to an observer moving with the same velocity as the particle.

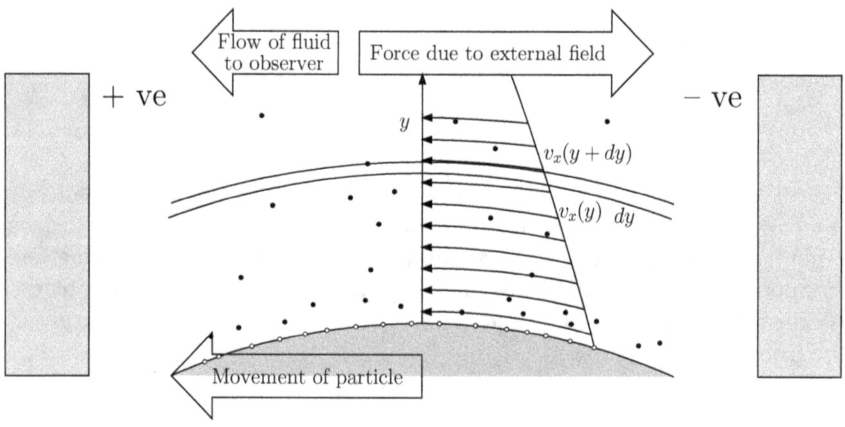

Figure 4.9 The situation in the EDL when $R << \kappa^{-1}$. The surface charge is negative.

two forces acting on them are viscous force due to fluid flow and electrostatic force due to the electrodes. As can be seen from Figure 4.9, they are acting in opposite directions and their magnitude must be equal.

To estimate the magnitude of the viscous force, consider the elementary volume shown in Figure 4.9. The x direction shear stress is given by

$$\tau_{xy} = -\mu \frac{dv_x(y)}{dy}.$$

If the net x directional viscous force per unit volume is denoted F_v,

$$F_v dxdydz = -\mu \left.\frac{dv_x(y)}{dy}\right|_{y+dy} dxdz + \mu \left.\frac{dv_x(y)}{dy}\right|_{y} dxdz,$$

which gives

$$F_v = -\mu \frac{d^2v}{dy^2}.$$

Balancing the viscous force with the electrostatic force:

$$\mu \frac{d^2v}{dy^2} = E\rho = E\epsilon \frac{d^2\psi}{dy^2},$$

where E is the external electric field. Integrating once:

$$\mu \frac{dv}{dy} = E\epsilon \frac{d\psi}{dy} + C.$$

At a large distance from surface, both dv/dy and $d\psi/dy$ become 0. Hence $C = 0$, which gives

$$\mu dv = E\epsilon d\psi.$$

Now integrating between limits:

$$\mu \int_0^u dv = \epsilon E \int_0^\zeta d\psi$$

or

$$\zeta = \frac{\mu u}{\epsilon E} = \frac{\mu(u/E)}{\epsilon},$$

where u is the observed velocity of the particle. In actual measurements, the velocity of the particle is observed using a microscope if the particle is sufficiently large. For submicron particles, light-scattering techniques are used.

4.7 Exercises

Exercises 4.1 Uniform spherical silver particles ($R = 100$ nm) are suspended in 12 mM aqueous AgBr solution at 298 K. The surface potential is measured to be 50 mV. The Hamaker constant for the system is 8×10^{-20} J.

- Calculate the Debye length of the system.
- If the net potential is written as

  $$\frac{\Phi_{net}}{kT} = c_1 \exp(-\kappa r) - \frac{c_2}{r}$$

  evaluate the constants c_1 and c_2 for this system.
- Evaluate and plot $\frac{\Phi_{net}}{kT}$ versus r in the range of 0.5 nm to 10 nm.

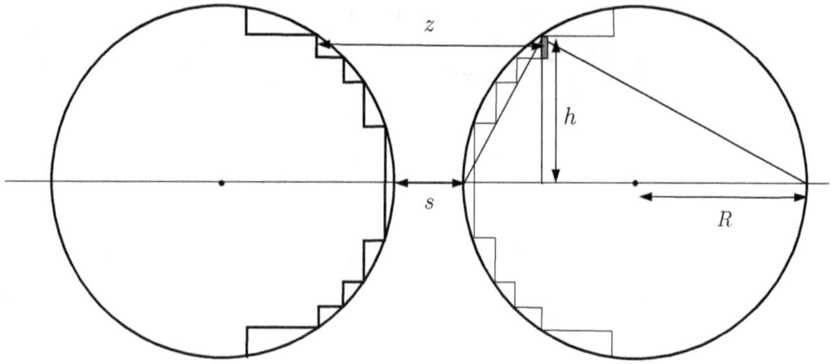

Figure 4.10 Interaction between two spheres.

- Determine the maximum energy barrier and the corresponding separation distance. Estimate the approximate stability ratio given by

$$W = \frac{\sqrt{\pi}}{4Rp} \exp\left(\frac{\Phi_{max}}{kT}\right)$$

where

$$p^2 = -\frac{1}{2kT}\left(\frac{d^2\Phi_{net}}{dr^2}\right)_{max}.$$

- Estimate the half-life of the colloid.

Exercises 4.2 The repulsive interaction energy (per unit area) between two infinite flat surfaces in an electrolyte solution separated by a distance z is given by the equation just above eqn 4.10.

a) Analyze the distance dependance of the interaction energy and estimate the change in interaction energy if the gap between the plates increases from 5 nm to 50 nm. You may assume properties typical to an aqueous system as given in the previous problem.
b) Consider two spheres, as shown in Figure 4.10, where $s << R$. Approximate the spheres as a stack of flat rings as shown in the figure and show that the sphere–sphere repulsive interaction energy is given by

$$\frac{\Phi_R(s)}{kT} = \frac{64\pi R\rho_0\gamma_0^2}{\kappa^2} \exp(-\kappa s).$$

It may be safely assumed that stacks corresponding to small h make a major contribution to the potential. Also note that h and z are not independent.

Exercises 4.3 Is ζ potential the true surface potential?

Bibliography

1 D. Myers. *Surfaces, Interfaces and Colloids: Principles and Applications*. VCH Publishers, 1991.

2 P. C. Hiemenz. *Principles of Colloid and Surface Chemistry*, 2nd edition. Marcel Dekker, 1986.

3 R. J. Hunter. *Foundations of Colloid Science*, 2nd edition. Oxford University Press, 2001.

4 J. N. Israelachvili. *Intermolecular and Surface Forces*, 3rd edition. Academic Press, 2011.

5 D. Fenell Evans and H. Wennerström. *The Colloidal Domain: Where Physics, Chemistry, Biology and Technology Meet*, 2nd edition. Wiley-VCH, 1999.

6 A. W. Adamson and A. P. Gast. *Physical Chemistry of Surfaces*, 6th edition. John Wiley & Sons, 1997.

5

Elementary Concepts of Number Balance

In the previous chapters we discussed three fundamental processes, namely, nucleation, growth, and aggregation. We did not refer to any particular equipment for their discussion and the models are generally valid. In the rest of the book we will develop models for equipments that deal with particulate systems. Since the equipment contains a collection of particles of varying size and shape we need to consider a population of particles.

A variety of systems can be described as a collection of particles. Such systems may be as abstract as a human population or as direct as a crystallizer. In chemical engineering, almost every process involves a particulate system at some stage, for example a mixer-settler in liquid–liquid extraction, rising bubbles on a distillation column, boiling liquid near a heat transfer surface, polymerization, solid catalytic reactors, fluidized beds, and granular flow.

We shall consider the crystallizer as a representative example to introduce the main concepts of number balance, which is the subject of this chapter. Let us consider a mixed suspension mixed product removal (MSMPR) crystallizer, as shown in Figure 5.1. A MSMPR crystallizer is similar to a continuous stirred tank reactor (CSTR) in many ways [1, 2]. Like a CSTR, the vessel is cooled using a jacket or cooling coil and remains well mixed so that the exit particle size distribution is same as that inside the vessel. The feed to the MSMPR crystallizer is a saturated or supersaturated solution which may or may not contain particles. Because of the cooling, particle nucleation and growth occur inside the vessel along with breakage and aggregation. The resulting slurry leaves the vessel with the exit stream.

Various special cases of MSMPR crystallizer are of importance. If the feed is free of particles, this crystallizer is called an unseeded crystallizer. For the unseeded case, nucleation and growth of particle occur inside the crystallizer and the outgoing stream contains particles. If the inlet contains seed, the seed particles grow in size but additional particles may also form due to primary or secondary nucleation. In absence of inlet or outlet streams, the crystallizer becomes a batch crystallizer.

Engineering of Submicron Particles: Fundamental Concepts and Models, First Edition. Jayanta Chakraborty.

Companion website: http://booksupport.wiley.com

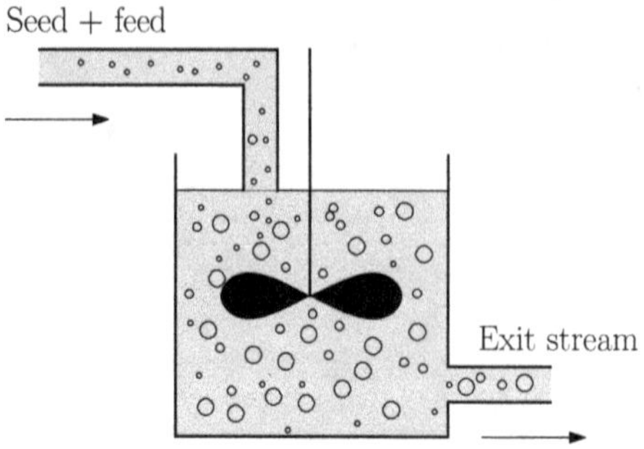

Figure 5.1 An MSMPR and a batch crystallizer. In both cases, the vessel is well mixed.

Our goal is to predict the particle size distribution at the exit of the crystallizer as a function of operating conditions such as cooling, flow rates, and feed composition. We will assume that particles are small enough so that they remain dispersed in the flow. For larger particles other considerations may be important [3]. Hence, we need to have a suitable description of the temporal state of the system, i.e. particle size distribution. We will start with another look at the description of a single particle.

5.1 State of a Particle

As we have seen before, a particle can be identified by its size. We have also seen that for faceted crystals, a single size variable is not enough to describe the crystal. Hence, one or more variables are required to describe a particle. The set of variables required to identify a particle uniquely are called its state variables. There is often more than one state variable and they are written as a vector called the particle state vector.

Size and shape are only two of the many attributes that are required to describe a particle uniquely. Consider the case where drops of pure benzene are mixed with drops of pure toluene in a well-stirred vessel. Due to the fluid movement, benzene drops coalesce with toluene drops. Larger drops will also break into smaller drops because of fluid shear. Hence, most of the drops will become a mixture of benzene and toluene. Now, although drops will be close to spherical, specifying the diameter (d) alone will not specify the drop completely. Drops with the same diameter may have different compositions (x). Hence, to describe a drop uniquely for this system, we must specify the mass fraction of benzene (or toluene) along with the diameter. This will complete the description and any two drops with the same diameter and mass fraction of benzene are indistinguishable. The particle state vector for this case may be written as $[d \quad x]$.

In many cases, a large number of state variables are needed to describe the particle. Consider the case of the viral infection of insect cells where insect cells are considered as particulate species floating in a nutrient solution containing virus. The number of viruses in a cell (n_v) is, of course, a state variable, but cell mass (m) is also important. Viruses multiply within a cell and during such multiplication they produce potent (vp) as well as defective (vd) viruses. Cell age (τ) is also important because after a certain age the cell will die and release viruses. Hence, the state vector becomes $[n_{vp} \quad n_{vd} \quad m \quad \tau]$. For a more refined model, many more state variables, such as the concentration of several proteins inside the cell, are also required.

The state variables discussed above are something inherent to the particle. No matter where the particles go, they carry that identity. For this reason, the state variables are called internal state variables or internal coordinates. Another type of state variable is location. Since the location of a particle is not something inherent to the particle and is dependent on external conditions such as flow, it is called an external state variable or external coordinates.

An external state variable is the coordinate of a particle with respect to a fixed origin and is denoted by the vector $\mathbf{r}$. The external coordinate is required when the system is spatially non-uniform. Particles of the same internal co-ordinate may behave differently depending on their location, i.e. external coordinates. For example, during cooling crystallization, if the crystallizer has different degrees of cooling at different locations, particles of the same size will have different growth rates at different locations. The complete description of a particle is therefore given by a combined vector of internal and external coordinates, called the particle state vector.

Example 5.1 Define suitable internal coordinates for the following situations:

1) Copper sulphate crystals suspended in a saturated copper sulphate solution, as shown in Figure 5.2.
2) Copper nanoparticles suspended in an aqueous medium, as shown in Figure 5.3.
3) Bacterial cells in a culture medium.
4) Air bubbles in a froth flotation tank.

Solution: As evident from Figure 5.2, the crystals are irregularly shaped. Hence, one measure of identity could be the mass of the crystal. If more elaborate description is needed, the internal coordinate vector will be the $\mathbf{h}$ vector, as described in section 2.2.

The TEM picture (Figure 5.3) shows that the particles are nearly spherical. Hence, particle diameter is a good choice for the internal coordinate.

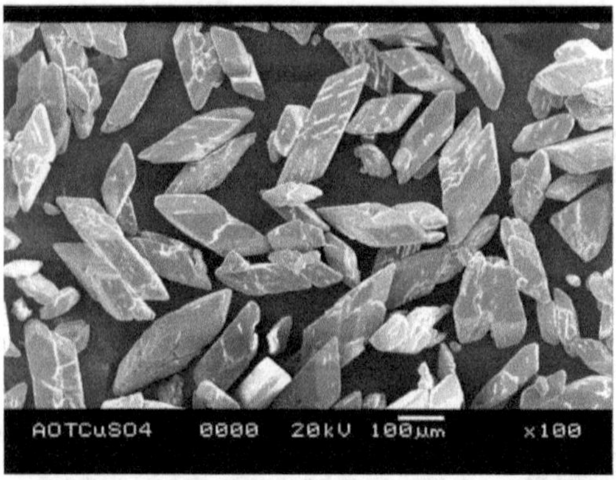

Figure 5.2 SEM micrograph of copper sulphate pentahydrate crystal.

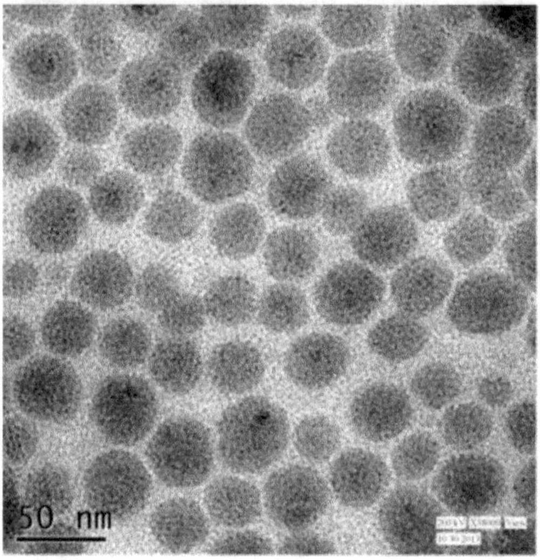

Figure 5.3 TEM micrograph of copper nanoparticles.

For bacterial cells in a culture medium, cell mass is a good choice for the internal coordinate. Cell age may be an additional variable.

Air bubbles in a froth floatation tank contain adsorbed mineral particles on their surface. Hence, the bubble size as well as the amount of minerals adsorbed on it are the two important internal coordinates for the bubble. □

Example 5.2 Determine whether or not external coordinates are required in the following situations:

1) Copper sulphate crystals and a supersaturated solution are fed to a long pipe for cooling.
2) Copper nanoparticles suspended in an ideal CSTR.
3) Particles freely settling in a stagnant liquid.

Solution:

1) For flow through a long pipe, the supersaturation is dependent on the location because the supersaturation depletes along the length. Hence, external coordinates are required.
2) Spatial variation is not possible in an ideal CSTR. Hence, external coordinates are not required.
3) Freely settling particles collide and possibly aggregate with other particles on their path. Hence the location of the particles is important. □

Particles are usually suspended in a fluid. The suspending fluid is called the continuous phase. For example, particles in a crystallizer are suspended in the mother liquor. In this case, the mother liquor is the continuous phase. The cells are suspended in a nutrient medium. The nutrient medium is the continuous phase in this case. The condition of the mother liquor/nutrient medium determines the growth of the particle phase and hence the properties of the continuous phase are important variables.

More than one property of the continuous phase might be of interest for the case in hand. Hence, in general, the relevant properties of the continuous phase are written in the form of a vector called the continuous phase vector. The continuous phase vector is denoted by the symbol $\mathbf{y}$. For example, in a crystallizer the degree of supersaturation may be the only property of the continuous phase that is of importance. For growth of cells in a culture medium concentrations of multiple nutrients may form the continuous phase vector.

5.2 State of a Population of Particles

In the previous section we discussed how we could describe a single particle precisely. Now we shall discuss what is a good description of a population of particles. There are at least three ways in which we can describe a population.

We will take the easiest and most tedious one first. Tag each and every particle at a given instant and give them some unique identification number (like each student has an enrollment number). Make a table for the values of their internal and external coordinates at each instant. It can be seen that taking such detailed descriptions is applicable only for small populations. Universities maintain records for their student population using such descriptions and many government organizations do it for a fairly large population. For chemical engineering systems, however, such a description is practicable only when the system is very small or a small portion of the system represents the actual system accurately.

Example 5.3 Mathematically represent a carrom board using the description discussed above.

Solution: There is a total of 20 particles, including one striker. The internal coordinates of the particles are size and colour. The location of the particles is important because depending on the board's position the player selects the move. Hence, the system may be characterized by the following table:

Tag	Size (cm)	Colour	*x* (m)	*y* (m)
1	2	Black	0.5	0.5
2	2	Black	0.1	0.05
3	2	White	0.1	0.5
⋮	⋮	⋮	⋮	⋮
20	3	Blue	0.2	0.4

□

Note that in the above example a variety of internal coordinates is used. Internal coordinates may or may not have a numerical value. If they do have a numerical value, it can be discrete (for this case) or continuous (for drop size or cell mass case).

This representation of a population of particles is not practicable for chemical engineering systems because such systems contain very large number of particles. For example, a crystallizer may contain 10^{10}–10^{13} particles per cubic centimetre. Also characterization is required on a daily or hourly basis. Hence, a coarser approach is taken. Instead of measuring the internal coordinate of each particle, the number (or mass) of particles in a given range of an internal coordinate is measured. Usually, a fixed amount of sample is taken (e.g. 1 L of slurry from a crystallizer or 1 kg of powder from the grinding product) and the number (or mass) of particles in a range of internal coordinates (say, particle diameter) is measured.

A familiar example of such characterization is sieving. Consider the characterization of a powder sample of spherical glass beads ranging from a few micron to 1 mm obtained from fumed silica processing. To analyze this sample, we can stack a series of sieves, the pan at the bottom and the coarsest sieve on the top, and place the powder on the topmost sieve and shake using a sieve shaker. We then measure the mass of powder collected on each sieve and plot the mass versus mean size. The representative size of the particles is taken as the arithmetic mean of the opening of the sieve above and below. The particle population is then presented using a histogram, as shown in Figure 5.4.

Example 5.4 How can you obtain the number distribution from the mass distribution as given above?

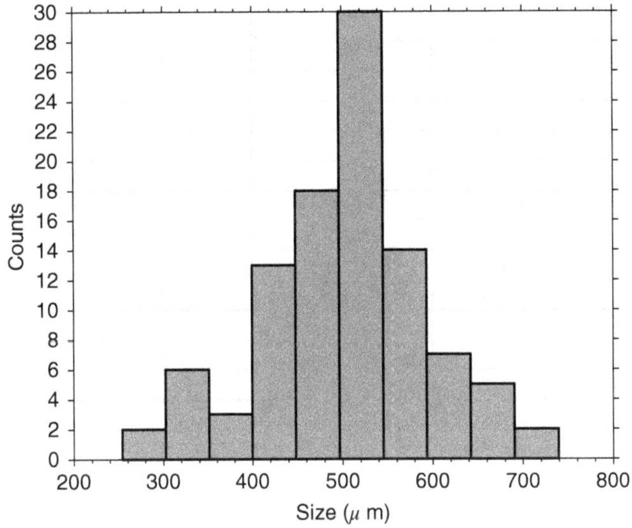

Figure 5.4 Representation of a population of particles using sieve analysis and a histogram.

Solution: This can be readily done by assuming that all particles in a size class are of equal size. If their number is denoted as N_i, the relation between mass and number becomes

$$m_i = (4/3)\pi r_i^3 \rho N_i. \qquad \square$$

While such a representation of a population of particles is widely used, the description is not very accurate. The inaccuracy emanates from the fact that particles over a range of size are represented by a single size and hence high inaccuracy results if the bin is wide. Narrowing down the bin reduces the error but this has a practical limitation. Usually, the amount of sample that can be processed quickly is small and hence smaller bin size leads to fewer particles in each bin, giving statistical error. It turns out that this is also a conceptual limitation restricting the use of the histogram as a good mathematical description of the particulate system.

On careful reflection, it can be seen that we cannot shrink the bin to an arbitrarily small size because as the bin shrinks, the number of particles in it also reduces and as the bin size approaches zero, the number of particles in it also approaches zero. This is because we are trying to define an extensive property as a point function. The number of particles in a size range increases as the width of the range increases and hence behaves like an extensive property in bin width. This is analogous to mass being an extensive property in space. Because a point, by definition has zero volume, it cannot hold any mass. Similarly, a point on the size axis has no 'volume' and hence cannot have any particle in it.

This problem is not unique to particulate systems. In all continuum descriptions of transport processes, mass balance is expressed using density instead of mass. Although the mass of a substance is zero at a given point, density can be defined as a point function. In a similar fashion, we can define an intensive variable instead of the number of particles in a size range.

However, the quantity used in the histogram, namely 'the number of particles in a certain range of sizes per unit volume of crystallizer' is already an intensive variable. It is and it is not. The number of particles per unit volume of crystallizer is an intensive variable with respect to the crystallizer volume but will shrink and expand with the width of the size range. Hence, to make it intensive with respect to the particle size coordinate, we should divide it by the bin width.

Hence, we can define the number density of particles as 'the number of particles in a given space *and* size range divided by the volume of the space *and* the width of the size range'. As the bin size shrinks, the number density converges to a definite value. This is demonstrated in Figures 5.5 and 5.6.

From the above discussion, we can write the definition of number density in the following way: if N_x is the number of particles in a size range x to $x + \Delta x$ collected from a small volume ΔV in a crystallizer, the number density is defined as

$$f_1(x,t) = \frac{N_x}{\Delta x \Delta V}.$$

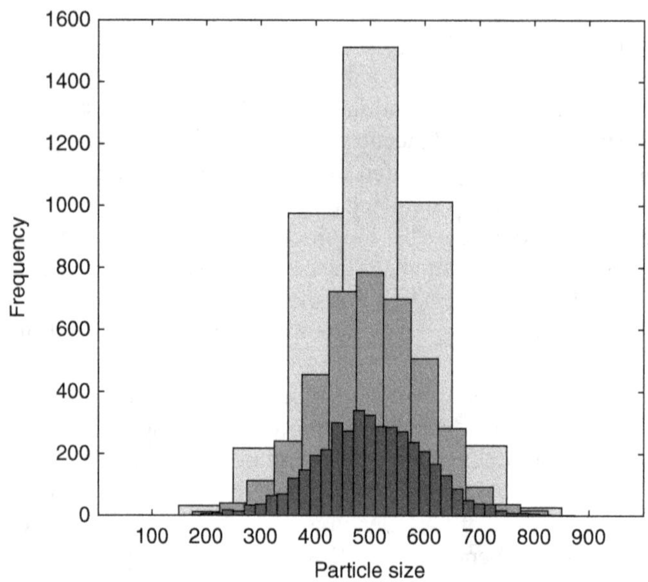

Figure 5.5 Particle size distribution of the same sample obtained with three different bin sizes. As the bin size becomes smaller, the number of particles in each bin reduces.

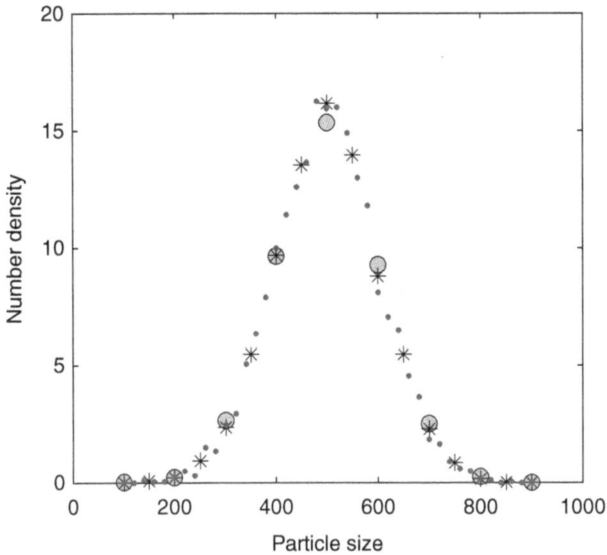

Figure 5.6 Number densities obtained by dividing the bin population by the bin width. All three bin widths approximately produce the same number density.

This relation can also be written for a differential size and volume range:

$$f_1(x,t)dx = \frac{N_x}{dV}.$$

The above illustration is deliberately oversimplified to display an important idea clearly. If external variables are important, the location of the elementary volume dV should also be mentioned. If the number density is defined with more than one internal coordinate, the differential range of each of the internal coordinates should be used. Hence, as a general case, the relation between number and number density becomes [4]

$$N(\mathbf{x},\mathbf{r}) = f_1(\mathbf{x},\mathbf{r})dV_{\mathbf{x}}dV_{\mathbf{r}},$$

where the differential volume in internal coordinate space ($dV_{\mathbf{x}}$) is simply the product of differential lengths in all required internal coordinates.

The reader may have an impression at this point that the description of a system of particles using a histogram is always an approximate one, but there are situations where such descriptions are accurate and we do not need to approximate a range of sizes by a single size. For example, consider a case of nano-sized emulsion droplets. Each emulsion droplet may contain at most 30 molecules of a certain species. Hence, if we describe the system by a histogram containing 30 classes, we provide an accurate description of the system. Such descriptions are also used frequently to characterize particulate systems.

Example 5.5 Select a proper description for the particulate systems for the following cases:

1) A large number of oligomers (1–10 primary units) in a solution.
2) A culture medium consisting of a few cells.
3) A polymer solution containing a very large number of polymers.

Solution:
1) Since oligomers have 1–10 monomeric units, they are not suitable for a continuous description. A histogram with 10 size classes will describe the system accurately.
2) A medium containing a few cells is also not suitable for continuous description. A table similar to that described in Example 5.3 will be more appropriate.
3) The internal coordinate may be the number of monomers, which ranges from a few to several thousand. A continuous description in terms of number density suits this case. □

Now we are ready to develop the number balance equation for a crystallizer. We shall start with a seeded batch crystallizer where no aggregation or breakage occurs. Particles will be assumed to be spherical so that only one internal coordinate is sufficient. The crystallizer is also assumed to be well mixed so that no spacial gradient exists and hence external coordinates are not important.

Developing the state equations for the MSMPR crystallizer requires us to write mass, energy, and momentum balances. In the particulate phase, the mass of particles is divided into infinitely many small classes. Hence, along with the total mass balances, it is possible to write a component mass balance where each 'component' is particles of a particular size. Usually, this component mass balance is written as a numbers balance. Hence, for a particulate system the available balance equations are total mass balance, number balances, momentum, and energy balances. The number balance equation is called the population balance equation (PBE).

5.3 Number Balance for a Seeded Batch Crystallizer

A seeded batch crystallizer is shown in Figure 5.1. A saturated solution and seeds are present at the beginning and at time $t = 0$ the temperature is lowered to generate the supersaturation. The crystals start to grow. No inlet or outlet stream exists, and aggregation and breakage are not significant under the conditions of the crystallizer. This is usually the case if the number of particles is low. For a well-mixed crystallizer with spherical particles, the number density function can be written as $f_1(x, t)$. The main phenomenon for the seeded batch crystallizer is growth of existing particles and hence we need a suitable growth model before we can develop the balance equations. If the growth is diffusion controlled and x is the diameter of the particles, the rate of growth is given by (eqn 2.1)

$$\frac{dx}{dt} \equiv G(x, c(t)) = \frac{4D}{\rho_p x}(c(t) - c_{sat})$$

Each of the particles in the crystallizer will grow according to this law and consume supersaturation.

Let us now write the number balance for the particles in the arbitrary range of sizes $[x_a, x_b]$, as shown in Figure 5.7. Note that this range is *not* a differential range. The number of particles per unit volume of crystallizer in this size range is given by $\int_{x_a}^{x_b} f_1(x, t)dx$ and hence the rate of change of number of particles in this range is

$$\frac{d}{dt}\int_{x_a}^{x_b} f_1(x, t)dx.$$

The number of particles in this region changes due to growth. The particles immediately left of boundary x_a will grow and will qualify (how many? not yet known) to be considered within the range $[x_a, x_b]$ and the particles which are just on the left side of the boundary at x_b will leave this range by growth. Hence,

$$\frac{d}{dt}\int_{a}^{b} f_1(x, t)dx = \dot{N}_L - \dot{N}_R,$$

where $\dot{N}_L$ and $\dot{N}_R$ are the number of particles that enter and leave through left and right boundaries, respectively, per unit time.

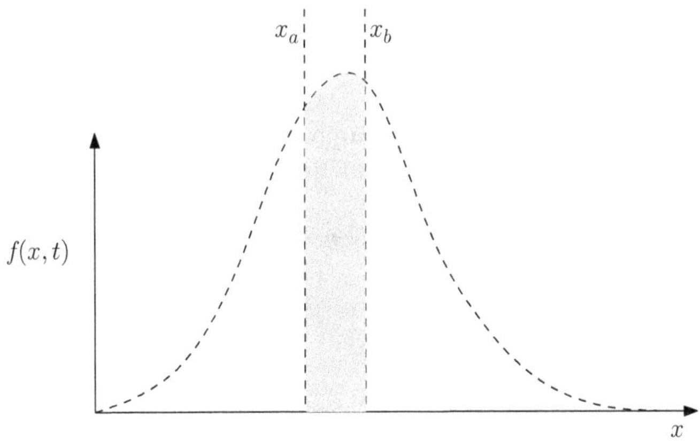

Figure 5.7 Number density function at an instant in a batch crystallizer. The shaded area represents the number of particles in the given range.

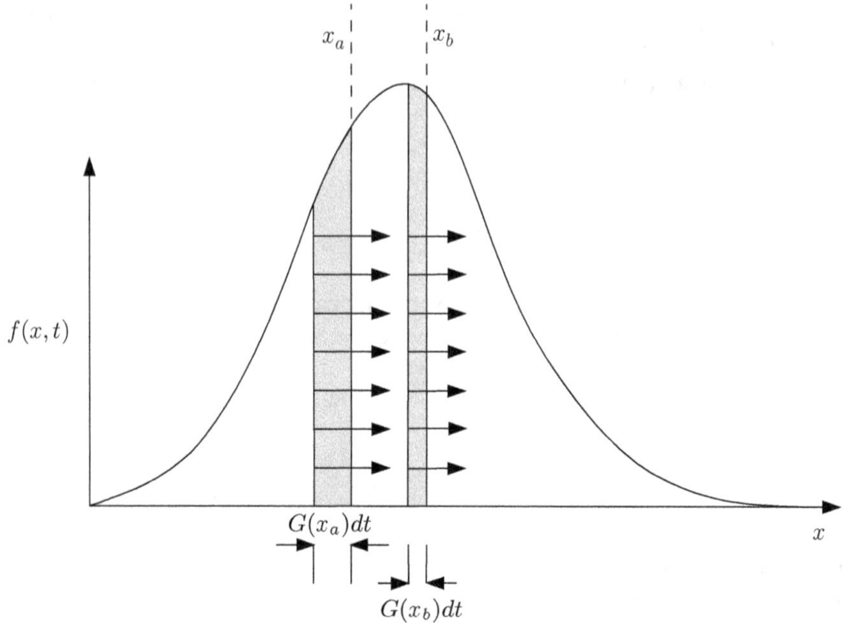

Figure 5.8 Number balance for a pure growth process. The number of particles that will enter and exit during the interval *dt* is marked.

Now consider a differential time interval dt. How many particles will enter the region $[x_a, x_b]$ during this time interval? It will surely depend on the growth rate. If the growth rate is $G(x, c(t))$, particles will grow by a size $G(x, c(t))dt$ during this interval and hence particles that are away from the boundary at x_a by at most a distance $G(x, c(t))dt|_{x_a}$ will be able to enter the region $[x_a, x_b]$ during this differential time interval dt. This range is shown as the shaded region in Figure 5.8 and all the particles in this region will enter the subdomain of interest in time dt.

As per the definition of number density, the number of particles in this small strip is $f_1(x, t)G(x, c(t))|_{x_a} dt$. Hence, the *rate* at which particles enter the domain of interest is $f_1(x, t)G(x, c(t))|_a$. Similarly, the rate at which particles leave the boundary b is $f_1(x, t)G(x, c(t))|_b$. Hence, the number balance can be written as

$$\frac{d}{dt}\int_a^b f_1(x,t)dx = f_1(x,t)G(x,c(t))|_a - f_1(x,t)G(x,c(t))|_b. \tag{5.1}$$

Figure 5.9 further clarifies this point.

The foregoing discussion is more general and useful than it appears at first. The rate at which particles enter through a boundary is given by the number density times growth rate. This is similar to transport flux at a boundary, which is given by concentration times velocity. In this case, the concentration is

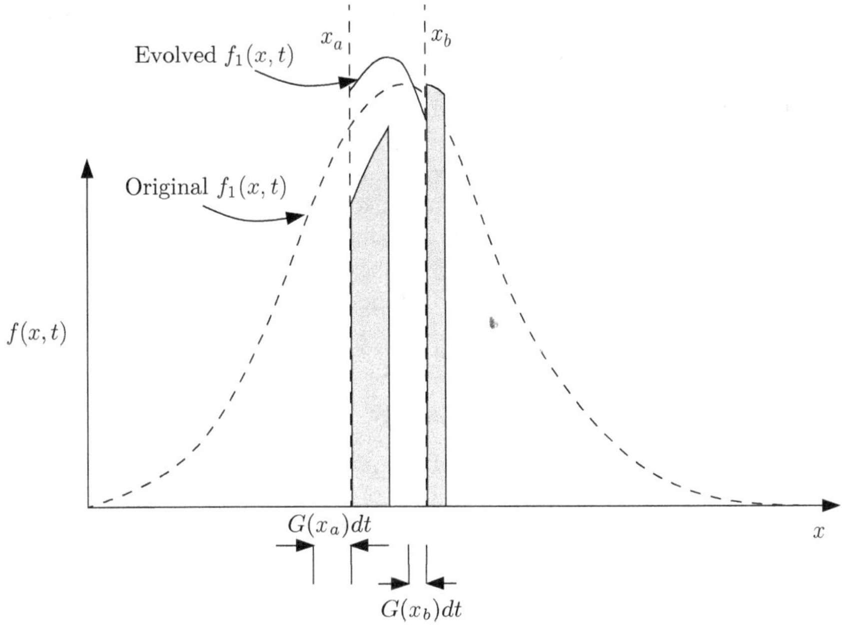

Figure 5.9 Number balance for a pure growth process. The evolved number density after the interval *dt* is marked.

replaced by number density and the velocity is replaced by growth rate, which can be thought of as velocity on the size axis.

Equation 5.1 is the desired number balance equation. Some mathematical clean-up is necessary at this stage: The right-hand side of eqn 5.1 can be modified and the equation can be written as

$$\int_a^b \frac{\partial f_1(x,t)dx}{\partial t} = -\int_a^b \frac{\partial}{\partial x}[G(x,c(t))f_1(x,t)]dx.$$

Bringing all terms to the left hand side gives

$$\int_a^b \left[\frac{\partial f_1(x,t)}{\partial t} + \frac{\partial}{\partial x}G(x,c(t))f_1(x,t)\right] dx = 0.$$

Because the interval $[a,b]$ is arbitrary, the quantity in the square bracket must be zero in order to make the integral vanish. Hence, the number balance equation finally becomes

$$\frac{\partial f_1(x,t)}{\partial t} + \frac{\partial}{\partial x}G(x,c(t))f_1(x,t) = 0. \tag{5.2}$$

This is the number balance equation for pure growth process. The model is not complete yet. This is a partial differential equation and hence proper initial and boundary conditions must be supplied. The initial condition is the number density of the seeds, which one can obtained from measurement. What about the boundary conditions? There can be various forms of boundary conditions for this equation depending on the situation. One of the possible boundary condition is $f_1(0,t) = 0$, which is equivalent to assuming that there is no 'zero size' particle in the crystallizer. This is a reasonable assumption if nucleation is absent. Now the PBE is complete but it cannot be solved yet because the transient concentration, $c(t)$, is an additional unknown. Hence we need another equation for $c(t)$ which is obtained from mass balance.

5.3.1 Coupling the PBE with Mass Balance

Because the system is a batch crystallizer, the total mass of solute remains constant. Initially it is mostly in the form of dissolved substance and a little amount as seed. Later, the size of the particles increases and the dissolved amount decreases. Let us denote $V_s(t)$ as the total volume of the solid phase, $V_l(t)$ as the total volume of the liquid phase, and $V_c(t)$ as the total slurry volume in the crystallizer. Hence, the mass balance can be written as

$$c(0)V_l(0) + V_c(0) \times m_s(0) \equiv M_T = c(t)V_l(t) + V_c(t)m_s(t),$$

where $m_s(t)$ is the mass of crystal per unit volume of slurry. Note that the quantity M_T is constant and it is also known from the initial conditions. Using the number density function,

$$m_s(t) = \rho_p v_s(t) = \rho_p \int_0^\infty \frac{\pi}{6} x^3 f_1(x,t)dx,$$

where $v_s(t)$ is the volume of particles per unit volume of slurry, the total slurry volume can be written as

$$V_c(t) = V_l(t) + v_s(t)V_c(t)$$

or

$$V_c(t) = \frac{V_l(t)}{1 - v_s(t)}.$$

Substituting these quantities in the mass balance equation gives

$$M_T = c(t)V_l(t) + \frac{V_l(t)}{1 - \int_0^\infty \frac{\pi}{6} x^3 f_1(x,t)dx} \rho_p \int_0^\infty \frac{\pi}{6} x^3 f_1(x,t)dx.$$

This set still has one unknown, $V_l(t)$. We need to consider the volume change of dissolution to obtain this quantity and in principle $V_l(t)$ may be obtained as a function of $c(t)$. For many cases, this complication can be avoided by assuming the total slurry volume to be constant. In that case the transient concentration is given by

$$c(t) = \frac{M_T - V_c \rho_p v_s(t)}{V_c(1 - v_s(t))}.$$

5.3.2 Modification for the Unseeded Case

How does our model equations change if nucleation and growth both occur in the crystallizer? First, we note that the nuclei are very small and may be thought of as zero size particles for all practical purposes. The appearance of particles of zero size happens only at a boundary of the domain and hence does not affect the differential equation. However, the events at the boundary will appear as a boundary condition. Hence, nucleation will change the boundary condition at $x = 0$.

As soon as a zero size particle is born through nucleation, the particle grows to a non-zero size and enters the domain. Hence, the rate of birth through nucleation should be exactly the same as the rate at which particles enter the domain through the boundary at $x = 0$. As discussed before, the rate at which particles enter any part of the distribution is a product of the growth rate and number density at that location. Hence, the following boundary condition can be written for this case:

$$G(0, t) f_1(0, t) = \dot{N},$$

where $\dot{N}$ is the nucleation rate. ($\dot{N}$) is a function of concentration and its expression for homogeneous nucleation is given in section 1.2.2. Although $G(0, t)$ is not known, it may be assumed to be same as the growth rate of the smallest measurable particle.

5.4 Number Balance for Open Systems

So far our concern has been a batch system. What modifications are needed to consider the inflow and outflow of streams? In this case, it is easier to write the material and number balances for the entire crystallizer rather than on a per unit volume basis. First, let us write this balance for the batch system by multiplying eqn 5.1 by the total volume of the slurry:

$$\frac{d}{dt}\int_a^b f_1(x, t) dx V_c(t) = f_1(x, t) G(x, c(t)) V_c(t)|_a - f_1(x, t) G(x, c(t)) V_c(t)|_b. \tag{5.3}$$

Now we can include inflow and outflow terms in the above equation. If the particles entering the vessel have a number density $f_{1,in}$ and the volumetric

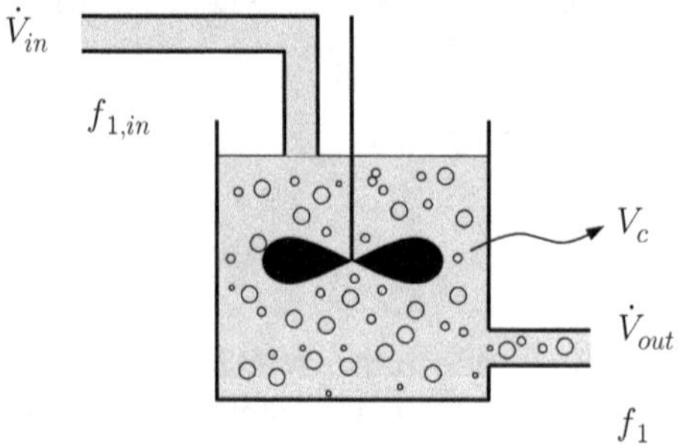

Figure 5.10 An MSMPR crystallizer.

flow rate of the entering slurry is $\dot{V}_{in}$, the number of particles entering in the size range of interest is $\int_a^b f_{1,in}(x)dx\dot{V}_{in}$ and the number leaving the reactor is $\int_a^b f_1(x,t)dx\dot{V}_{out}$. Here an ideal mixed vessel is assumed so that the exit number density and the number density inside the vessel are identical. The number balance then becomes

$$\begin{aligned}\frac{d}{dt}\int_a^b f_1(x,t)dxV_c(t) = &f_1(x,t)G(x,c(t))V_c(t)|_a\\ &-f_1(x,t)G(x,c(t))V_c(t)|_b\\ &+\int_a^b f_{1,in}(x)dx\dot{V}_{in} - \int_a^b f_1(x,t)dx\dot{V}_{out}.\end{aligned} \tag{5.4}$$

Simplifying the equation in the same way as eqn 5.1 we obtain the number balance for the open system:

$$\frac{\partial}{\partial t}f_1(x,t)V_c(t) + V_c(t)\frac{\partial}{\partial x}G(x,c(t))f_1(x,t) = f_{1,in}(x)\dot{V}_{in} - f_1(x,t)\dot{V}_{out}. \tag{5.5}$$

Example 5.6 An MSMPR crystallizer is shown in Figure 5.10.

- If no seed is supplied and nucleation is happening through a homogeneous route inside the crystallizer, obtain the closed set of model equations for prediction of the number density for this crystallizer.
- Reduce the model for steady state operation of the crystallizer.
- Use McCabe ΔL law to simplify the steady state model derived above.
- Obtain the steady state number density function by solving the model.

Solution: Since no particle is present in the inlet stream, the number balance equation becomes

$$\frac{\partial}{\partial t}[f_1(x,t)V_c(t)] + V_c(t)\frac{\partial}{\partial x}[G(x,c(t))f_1(x,t)] = -f_1(x,t)\dot{V}_{out}.$$

Nucleation will serve as the boundary condition for this equation and the initial particle size distribution should be known, which will supply the initial condition:

$$f_1(x,0) = f_{10}$$
$$G(0,c(t))f_1(0,t) = k_{n1}\exp\left[\frac{k_{n2}}{-[\ln c(t)/c_{sat}]^2}\right].$$

We will also need a transient mass balance for this system. In general, nucleation consumes a very small amount of mass and can be excluded from the mass balance. Hence, the mass balance becomes

$$\frac{d}{dt}[V_l(t)c(t) + V_c(t)m_s(t))] = c_{in}\dot{V}_{in} - c(t)\dot{V}_{out}(1 - \nu_s(t)) - m_s\dot{V}_{out}.$$

Now let us check whether or not the model equations form a closed set. We can see that the PBE is complete with initial and boundary conditions provided the $c(t)$ is known. The mass balance can be solved for $c(t)$ if $V_l(t)$, $V_c(t)$, and $\dot{V}_{out}(t)$ are available. These three are the degrees of freedom for the model and they should be specified or known in order to solve the model.

At steady state, the flow, slurry volume, and concentration all achieve their respective steady state values. Let us denote the steady outflow rate by $\dot{V}$, the steady state crystallizer volume by V_{cs}, and the steady state concentration by c_s. The number density also achieves its steady state value, which is denoted by removing its time dependence. The time derivative of the number density becomes zero and the partial derivative in x becomes the total derivative. With these changes in notation, the steady state PBE becomes

$$V_{cs}\frac{d}{dx}[G(x,c_s)f_1(x)] = -f_1(x)\dot{V}.$$

Denoting $V_{cs}/\dot{V} = \tau$,

$$\tau\frac{d}{dx}[G(x,c_s)f_1(x)] = -f_1(x).$$

Since time derivatives are not present, the initial condition will not be required. The boundary condition for the steady state becomes

$$G(0,c_s)f_1(0) = k_{n1}\exp\left[\frac{k_{n2}}{-[\ln(c_s/c_{sat})]^2}\right].$$

The steady state mass balance equation becomes

$$c_{in}\dot{V}_{in} - c_s\dot{V}(1 - \nu_{s,s}) - m_{s,s}\dot{V} = 0.$$

This equation can be solved for c_s provided $\dot{V}$ is known. A flow model can be written for the steady state outflow rate as

$$\dot{V} = k\sqrt{\frac{V_{cs}}{A}\rho_{sl}(c_s, m_{s,s})},$$

which forms the closed set.

If the growth rate is independent of particle size (the McCabe ΔL law), the PBE becomes

$$G(c_s)\tau \frac{d}{dx} f_1(x) = -f_1(x).$$

Since c_s is constant, the functional dependence is meaningless here. Removing this from G and solving,

$$f_1(x) = f_1(0)\exp\left[-\frac{x}{G\tau}\right]. \tag{5.6}$$

This relation can be used to obtain v_s and m_s at steady state and then substituted into the mass balance equation to obtain the steady state concentration c_s. Quantitative treatment using experimental data for such systems is available in [5]. □

5.5 Exercises

Exercises 5.1 Relate the *number density function* to physically measurable quantities for (i) a well-mixed vessel and (ii) a vessel with spatial variation. Define mass density in a similar manner to the way we defined number density. Relate it to a physically measurable quantity. What is the relation between number density and mass density?

Exercises 5.2 Obtain the expression for the total mass of crystals per unit volume of slurry (usually termed magma density, M_T) as a function of growth rate (G) and residence time (τ) for steady state operation of an unseeded MSMPR under constant growth rate.

Exercises 5.3 Consider the growth of small perfectly spherical particles. The growth rate is not known and may be a function of particle size. If we consider a pair of particles whose diameters are d_1 and d_2, and at a given instant $d_1 > d_2$, is it possible that at a later time $d_1 < d_2$? Explain.

Exercises 5.4 Often the width of the particle size distribution is a key property that needs to be controlled in the operation of a crystallizer. The width is quantified by the ratio of the standard deviation and the mean of the particle size distribution, which is known as the coefficient of variance. Consider a seeded batch cooling crystallizer and comment on the evolution of the coefficient of variance for the following growth rates: (i) $G(r) = \frac{k}{r}$, (ii) $G(r) = k$, and (iii) $G(r) = kr^2$ where r is the particle radius and k is a function of supersaturation only.

Exercises 5.5 For an MSMPR crystallizer, the mass cumulative undersize, i.e. the total mass of particles up to a size l (denoted by$M(l)$), is an important quantity. For the case of an unseeded MSMPR crystallizer show that

$$\frac{M(x)}{M_{tot}} = 1 - \left(1 + x + \frac{x^2}{2} + \frac{x^3}{6}\right) e^{-x},$$

where $x = l/(G\tau)$. Evaluate the median size and suggest an easy way to control this parameter.

Exercises 5.6 The mass density (similar to the number density but instead of counting the number, we measure the mass) for an MSMPR crystallizer can be obtained from the cumulative relation obtained in Exercise 5.5. Show that the mass density function is given by

$$f_m(x) = \frac{x^3}{6} e^{-x}$$

and compare it with the number density function. Discuss the possible reasons for the differences, if any.

Exercises 5.7 The mode of the distribution $f_m(x)$ is an important quality control parameter. As a process control engineer, suggest a suitable manipulated variable to control this parameter. Your answer must be supported by clear mathematical deduction(s).

Exercises 5.8 In many cases nuclei are formed over a size range (still small, but cannot be approximated as zero size) instead of zero size particles. Develop a model for a batch cooling crystallizer to take this effect into account.

Exercises 5.9 During operation of an MSMPR crystallizer, the inlet concentration suddenly doubles. What will be the effect on the average particle size? Explain using a model.

Exercises 5.10 Consider the growth of cells in a culture medium. What will the growth rate be if cell age is the internal coordinate? Consider cell mass and cell age as two elements of the internal state vector. Derive the growth PBE for this case.

Exercises 5.11 A yeast cell gives birth to new yeast cells by forming a 'bud' on its surface. The bud grows and after a certain size leaves the mother cell, leaving a 'scar' on the cell surface. Then a new bud grows. As the number of scars on the cell surface increases, the 'budding' tendency decreases and finally the cell becomes impaired. Chose a suitable internal coordinate vector for this system.

Exercises 5.12 Defend or criticise the following statement: 'A population balance equation may be viewed as component mass balance'.

Bibliography

1 J. W. Mullin. *Crystallization*, 4th edition. Butterworth-Heinemann, 2001.
2 A. D. Randolph and M. A. Larson. *Theory of Particulate Processes: Analysis and Techniques of Continuous Crystallization*, 2nd edition. Academic Press, 1988.
3 M. J. Rhodes. *Introduction to Particle Technology*, 2nd edition. John Wiley & Sons, 2008.
4 D. Ramkrishna. *Population Balances: Theory and Applications to Particulate Systems in Engineering*. Academic Press, 2000.
5 A. S. Myerson. *Handbook of Industrial Crystallization*, 2nd edition. Butterworth-Heinemann, 2002.

6

Breakage and Aggregation

In the previous chapter we modeled a crystallizer where nucleation and growth occur. In this chapter we will incorporate breakage and aggregation into our population balance model.

There is a fundamental difference between particle growth and breakage/aggregation. During growth, the internal coordinate of a particle (volume, diameter, denoted by x) changes continuously. In other words, a particle moves along the internal coordinate in a continuous fashion. On the other hand, the internal coordinate of particles change by a 'jump' if breakage takes place.

For example, if a spherical drop of volume x breaks into two equal drops, the volume of the daughter particles changes to $x/2$. In fact, the original particle vanishes by giving birth to two particles with internal coordinates $x/2$ as if the mother particle 'jumped' out from its location in the internal coordinate axis to give two particles of internal coordinates $x/2$, which jump into their respective locations.

A similar feature is observed for the aggregation process. If two particles of volume $x/2$ aggregate, a particle of size x is formed. Hence, the change in internal coordinate corresponding to aggregation happens in a jump from $x/2$ to x. In this chapter we will discuss how to incorporate breakage and aggregation in our number balance equation. We will start with a brief discussion of the phenomenology of the breakage process.

6.1 Breakage Functions

During breakage, a 'particle' produces two or more particles. Breakage need not always be something like shattering a piece of glass but may be as intricate as the division of living cells. It can also be polymer degradation, attrition of large particles into fine micron size powder etc.

It can be seen that the physics of all these processes are widely different. For example, the breakage of cells is driven by a number of factors, such as cell age,

Engineering of Submicron Particles: Fundamental Concepts and Models, First Edition. Jayanta Chakraborty.

Companion website: http://booksupport.wiley.com

nutrient concentration etc. On the other hand, polymer degradation may happen due to high temperature or even the presence of UV light. Solid particles break due to propagation of cracks under applied forces. Despite such wide differences, it is possible to model breakage processes using three generalized functions. The functions that contain the physics of the breakage process are known as 'breakage functions'. There are three breakage functions that define a breakage process completely:

- **The frequency of breakage,** $\Gamma(x)$: The fraction of existing particles (in the range x to $x + dx$) that breaks per unit time.
- **The average number of daughter particles formed due to breakage of a particle of size** x, $\nu(x)$: This *average* number can be a fraction. For example, for breakage of a 300 μm particle, sometimes two daughter particles are produced and sometimes three particles are produced. If observed for a sufficiently large number of such events, it may be seen that in 50% of cases two particles are formed. Hence, the average number of daughter particles for this case is 2.5.
- **The daughter size distribution,** $P(x|x')$: The *normalized probability density function* for the production of particles of size x from the break-up of a particle of size x'.

It can be seen that the detailed physics of the breakage process should be known to specify the breakage functions accurately. However, in many cases only limited information is available regarding the process and a semi-empirical approach is used. Let us consider the degradation of polymers as an example. Polymers may degrade by a variety of processes. One interesting degradation route is shear induced polymer degradation. In this case, the polymer chains suspended in a solvent are caught in the shear field and pulled from both ends. Due to the symmetric nature of the pulling, the polymer molecules break near the mid-point. This kind of breakage is called mid-point chain scission.

The rate of breakage is dependent on the force of pulling. The longer the chain, stronger the pull. Hence the rate of breakage must increase with increasing chain length. If we take the molecular weight (x) as an internal coordinate for polymers, the chain length will be proportional to this variable. A simple power law or even linear dependence between the molecular weight and propensity to break may be assumed as a first approximation. Hence, we may write the breakage rate as

$$\Gamma(x) = kx.$$

This expression needs a small correction: if the chain length is too small, the 'pull' may not be sufficient to break the C–C bond. Hence, below a certain value of x the breakage rate must be zero. This may be incorporated in the model as

$$\Gamma(x) = \begin{cases} k(x - x_l) & \text{if } x > x_l \\ 0 & \text{if } x \leq x_l, \end{cases}$$

where x_l is the limiting molecular weight below which breakage is not possible. Although polymer degradation was taken as an example in the above, such models work well even for grinding of solids. Usually it is observed that the larger particles break more frequently compared to finer particles. This is because larger particles contain more micro cracks, although the number of micro cracks per unit volume is same for both small and large particles. As soon as one of the cracks gives way, the particle breaks. Also small particles are mostly obtained by breaking large particles. Hence, the weakest cracks are already broken. A grinding limit similar to the limiting molecular weight is therefore also observed for solid particles [1].

The daughter distribution function and average number of particles is readily determined in the above case if we assume that each chain breaks exactly from the mid-point (see Example 6.1). However, such logical determination is often very difficult and specification of breakage functions often needs careful experimentation. Even in the above case, the limiting molecular weight should be obtained from experiments. Breakage functions for solid particles in a grinding mill can also be obtained by observing the breakage of individual particles under representative conditions. For example, Vogel and Peukert [2] studied a large number of particles under impact to determine the breakage frequency. Collection of daughter particles and subsequent size analysis will lead to the other two breakage functions.

In many cases the linear breakage frequency is inadequate but a power law model explains the data adequately: $\Gamma(x) = kx^a$ where k and a are the two parameters of the model. These parameters may be related to the physics of the process such as the number of micro cracks per unit volume, but in most cases these are empirically fitted.

Example 6.1 Obtain the daughter distribution function for the following cases using inspection: breakage of polymer chains from (a) the mid-point, (b) the end, and (c) a random position.

Solution:

a) The process can be modeled as exact mid-point breakage and hence the daughter distribution function can be written as a Dirac-delta function:

$$P(x|x') = \delta(x - x'/2).$$

b) In this case the breakage also occurs at a specified location. Hence, the daughter distribution is again a Dirac-delta function, but the location is now just one unit away from the chain end. Hence, the daughter distribution is given by

$$P(x|x') = \delta(x - (x' - x_m)),$$

where x_m is the molecular weight of the monomer.

c) In this case the chain can break from anywhere, which means that the probability of breakage of any of the C–C bonds is equally likely. If the molecular weight of a polymer is x', then the polymer has x'/x_m bonds.[1] The probability of breakage of *any one of these* bonds is then x_m/x'. To obtain the probability density, we must divide the probability by width of the zone over which the probability is obtained. The width should be x_m, therefore

$$P(x|x') = 1/x'.$$

□

It is advisable that the breakage functions obtained are verified for their properties. For example, the daughter distribution function is a normalized probability distribution. Hence, it must obey the relation

$$\int_0^\infty P(x, x')dx = 1. \tag{6.1}$$

Another property of importance is that the size of the daughter particle produced must be less than that of the mother particle:

$$P(x, x') = 0 \qquad \text{If} \quad x > x'. \tag{6.2}$$

The third constraint that the daughter distribution must satisfy is the mass balance. The probability weighted mass of all possible daughter particles must be the same as the mass of the mother particle:

$$m(x') = \int_0^{x'} \nu(x')m(x)P(x, x')dx. \tag{6.3}$$

Here $m(x)$ is the mass of the particle with internal coordinate x. For binary breakage the following symmetry condition should also be satisfied:

$$P(x, x') = P(x - x', x'). \tag{6.4}$$

Example 6.2 It is proposed that the daughter distribution function for random breakage is $P(x|x') = x_m/x'$. Verify whether this daughter distribution function satisfies all the required constraints.

Solution: First, let us check the normality constraint:

$$\int_0^{x'} \frac{x_m}{x'} dx = x_m \neq 1.$$

Hence, the daughter distribution function needs normalization by x_m after which it becomes $1/x'$. The second constraint must be included in the definition and the fourth constraint is satisfied since it is not a function of x. Let us

1 Strictly speaking $x'/x_m - 1$, but since $x'/x_m >> 1$ it may be approximated to x'/x_m.

examine the third constraint. The mass of a polymer of molecular weight x' is $m(x') = x_m x'$:

$$\int_0^{x'} 2x_m x \frac{1}{x'} dx = \frac{2x_m}{x'} \frac{x^2}{2}\bigg|_0^{x'} = x' x_m.$$

Hence, all the constraints are satisfied by this corrected daughter distribution function. □

Often it becomes difficult to model the daughter distribution function exactly. For such cases, a probability density function consistent with the process may be used. One of the probability density functions that can take a variety of shapes depending on its parameters is Beta distribution. Its expression is given by

$$P(x|x') = \frac{x^{\alpha-1}(1-x)^{\beta-1}}{B(\alpha, \beta)},$$

where

$$B(\alpha, \beta) = \frac{\Gamma(\alpha)\Gamma(\beta)}{\Gamma(\alpha+\beta)}.$$

Here Γ is the gamma function. The Beta distribution functions for several values of α and β are shown in Figure 6.1. To model the attrition process, we may chose $\alpha = \beta = 0.5$.

Figure 6.1 Beta probability density function.

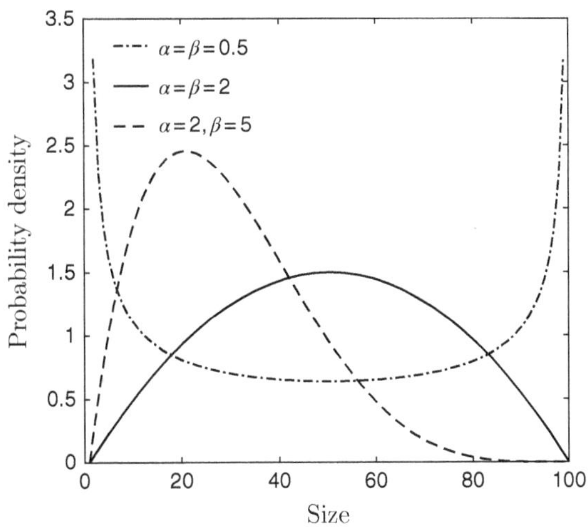

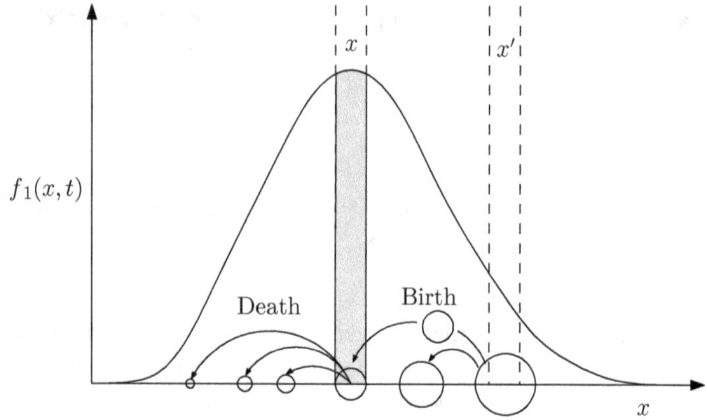

Figure 6.2 Birth and death due to breakage processes.

6.2 Number Balance for Breakage

Now we are ready to write the number balance for breakage process. To keep ourselves focused, we will assume that processes like growth, aggregation or flow streams are not present while writing the number balance for breakage. However, terms corresponding to these processes can be included in the number balance in a similar fashion, as discussed in the previous chapter.

We start by writing the number balance for particles residing in an elementary strip on the size axis, as shown in Figure 6.2. The change in the number of particles in this strip is due to breakage. If particles in this size range break, they become smaller (at least smaller by an amount such that they are no longer in this elementary strip) and hence do not qualify to be considered in this range any more. The rate at which such loss of particles occurs may be found by using breakage functions.

The function $\Gamma(x)$ gives the fraction of particles breaking per unit time. To find the number of particles breaking and leaving the elementary strip per unit time, the number of particles in the strip needs to be known. This is given by $f(x,t)dx$. Hence, the number of particles leaving the strip per unit time is given by $\Gamma(x)f(x,t)dx$. The process of loss is also called the death process.

Is this strip also gaining some particles? Surely some of the larger particles that are breaking will contribute into this size range, as shown in Figure 6.2. How many such particles form (rather jumping into) in the interval of interest per unit time? Let us consider another elementary strip at an arbitrary size x'.

The number of particles breaking per unit time from this strip is given by $\Gamma(x')f(x',t)dx'$. How many daughter particles are produced due to breakage of $\Gamma(x')f(x',t)dx'$ mother particles? This is obtained by multiplying $\Gamma(x')f(x',t)dx'$

by the second breakage function $\nu(x')$. Hence, the total number of particles produced per unit time from the breakage of particles in the elementary strip of thickness dx' around x' is $\nu(x')\Gamma(x')f(x', t)dx'$.

The next question is how many of *these* particles fall within the size range from x to $x + dx$? For this we require the third breakage function. Remember that $P(x|x')$ is a probability density function and hence $P(x|x')dx$ gives the probability (fraction) of daughter particles falling in the size range from x to $x + dx$ upon breakage of particles of size x' to $x' + dx'$. Hence, the number of daughter particles that fall in the size range from x to $x + dx$ upon breakage of particles around size x' is

$$\nu(x')\Gamma(x')f(x', t)dx'P(x|x')dx.$$

Now x' was chosen arbitrarily and any x' larger than x can produce particles in x. To account for this, we must sum all breakage birth terms by integrating with respect to x'. This will give the total gain of particles in the small strip around x due to breakage of larger particles:

$$\int_x^\infty \nu(x')P(x|x')dx\Gamma(x')f(x', t)dx'.$$

The last term is known as the breakage birth term. Now, combining both birth and death terms on the right-hand side and cancelling dx from both sides, the number balance becomes

$$\frac{\partial f_1(x,t)}{\partial t} = -\Gamma(x)f_1(x,t) + \int_x^\infty \nu(x')P(x|x')\Gamma(x')f(x', t)dx'. \tag{6.5}$$

Note that the number balance for breakage is an integro-partial-differential equation whereas the number balance for the pure growth process is a partial differential equation. The integral term is a characteristic of jump processes.

Example 6.3 An organic phase containing a water-soluble impurity is fed near the impeller of a mixing vessel for removing a water-soluble contaminant, as shown in Figure 6.3. Uniform drops are released by the dispenser and break into smaller drops in the mixer before leaving.

Using the following assumptions, develop the population balance equation to obtain the exit average concentration of impurity in the organic phase at steady state.[2]

1) The amount of dispersed phase is much lesser than of continuous phase, which means that:
   a) aggregation is negligible
   b) impurity concentration in the continuous phase remains negligibly small.

2 A more elaborate version of this problem is available in [3].

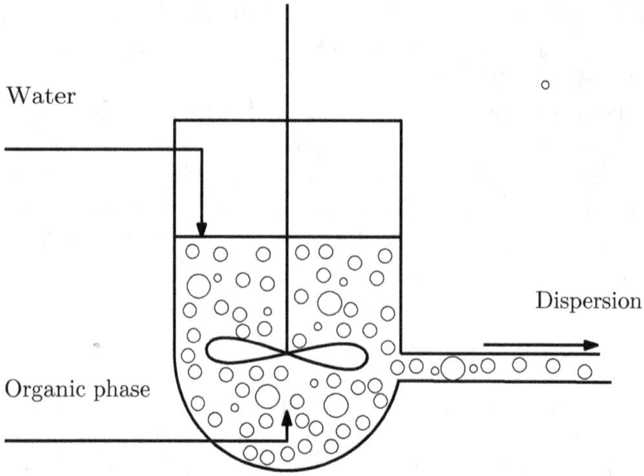

Figure 6.3 Liquid–liquid dispersion.

2) The process is mass transfer controlled, but the mass transfer resistance in both the phases is negligible except for a thin film near the surface.
3) Solute concentration does not influence the breakage of drops.
4) Mass transfer does not change the drop size significantly.

Solution: Let us first find a suitable internal coordinate vector. The particle size is required because particle breakage occurs. The total mass transfer area per unit volume of organic phase is dependent on the average particle diameter. Two particles of the same size may have different amounts of solute and hence solute concentration is also an internal coordinate. Hence, the number density can be written as $f_1(c, r, t)$.

Note the difference between the two internal coordinates used: the concentration coordinate changes gradually and in a way similar to particle size under growth process, but particle radius changes in a jump. In–out terms should also be included. Since the particles of largest size are introduced through the nozzle, corresponding terms should be considered in a boundary condition. Flow out terms will be treated in usual way. With these conditions, the population balance equation becomes:

$$\frac{\partial f_1(c,r,t)}{\partial t} + \frac{\partial}{\partial c}[\dot{c}(r)f_1(c,r,t)] = -\frac{f_1(c,r,t)}{V/\dot{V}_{out}} - \Gamma(r)f_1(c,r,t) + \int_r^{R_{Max}} \nu(r')P(r|r')\Gamma(r')f_1(c,r',t)dr'.$$

Here $\dot{c}$ is the rate of change of concentration of an individual drop of size r, which can be obtained using a suitable mass transfer model. The breakage terms can be readily understood. The out term is also similar to that developed previously.

For steady state the above equation becomes

$$\frac{\partial}{\partial c}[\dot{c}f_1(c,r)] = -\frac{f_1(c,r)}{(V/\dot{V}_{out})_{ss}} - \Gamma(r)f_1(c,r) + \int_r^\infty \nu(r')P(r|r')\Gamma(r')f_1(c,r')dr'.$$

In the above equation $(V/\dot{V}_{out})_{ss}$ signifies the steady state residence time. The boundary condition for the above equation can be written as

$$f_1(0,r) = 0.$$

The solute mass balance can be developed in a similar way, as shown in the previous chapter. □

6.2.1 Discrete Breakage Equation

For comminution (breakage) of solid particles, a discrete breakage equation is frequently used [1]. If M_i is the mass fraction of particles in the ith size class, the discrete breakage equation is given by

$$\frac{dM_i(t)}{dt} = -k_iM_i(t) + \sum_{i+1}^{N} k_j\nu_jM_j(t)B_{ij}. \tag{6.6}$$

The breakage functions used in this equation have similar meaning and empirical expressions for these breakage functions are readily available.

6.3 The Process of Aggregation

Aggregation is the process by which two or more particles collide and unify. In some cases, like liquid drops, the colliding particles loose their identity completely and the original particles cannot be identified in the aggregate. When solid particles aggregate due to attractive potential, the aggregate usually contains the grain boundary corresponding to the surface of the original particles, which may vanish later. In this chapter we will call both types aggregation and consider the aggregated particle as a single entity.

If two particles aggregate, this is called binary aggregation. Aggregation may be of higher order, such as ternary or quaternary. However, for dilute systems, binary aggregation is far more probable than higher order aggregations and in this chapter we will consider only binary aggregation assuming a dilute system.

Binary aggregation requires the formation of particle pairs. Hence, to model the process of aggregation we need to define the number density of particle pairs. Previously, we defined the number density of particles (singlets, to be precise) as a function of their internal and external coordinates. In a similar way, we can also define the number density of particle pairs as $f_2(x, x', \mathbf{r}, \mathbf{r}', t)$. The subscript 2 in the number density function signifies that this is the density of pairs and not singlets. Note that the internal coordinate has been assumed to be a scalar for the sake of simplicity. If nothing else is stated, the internal coordinate may be thought of as particle volume. Although we take these simplifications, the treatment that follows is general. Also note that we retain the external coordinates because the physical proximity of particles is required for aggregation. The quantity $f_2(x, x', \mathbf{r}, \mathbf{r}', t)dxdx'd\mathbf{V}_\mathbf{r}d\mathbf{V}_{\mathbf{r}'}$ represents the total number of pairs in which one of the particles lies in the size range x to $x + dx$ and the other lies in the size range x' to $x' + dx'$, and the first particle in the pair is located in the elementary volume $d\mathbf{V}_\mathbf{r}$ around $\mathbf{r}$ and second particle in the pair is located in the elementary volume $d\mathbf{V}_{\mathbf{r}'}$ at location $\mathbf{r}'$. This information may be obtained if snapshots of the system can be taken.

Example 6.4 Consider a carrom board just before the beginning of a game. Count the number of pairs where one of the pieces is red and the other is black, and the red piece is located at the centre of the board and the black piece is located at a distance D from its centre. The diameter of the pieces is D.

Solution: Three. □

Among $f_2(x, x', \mathbf{r}, \mathbf{r}', t)dxdx'd\mathbf{V}_\mathbf{r}d\mathbf{V}_{\mathbf{r}'}$ of existing pairs, how many will aggregate? One of the requirements for aggregation is proximity, which can be determined by looking at $\mathbf{r}$ and $\mathbf{r}'$. However, proximity alone does not guarantee aggregation. Some of the particle pairs may touch each other and then separate out. Usually a fraction of the particles that come into contact aggregate. This fraction is called the aggregation frequency. Usually the aggregation frequency is a function of factors such as temperature, pH, salinity of the continuous phase, particle size etc. Hence, the aggregation frequency can be written as

$$a(\mathbf{x}, \mathbf{x}', \mathbf{r}, \mathbf{r}', \mathbf{y}(\mathbf{t})).$$

Like breakage frequency, this aggregation frequency can be read as 'the fraction of pairs aggregating per unit time'.

Another important point we need to consider before writing the aggregation number balance is the state of the newly formed particle after aggregation. If the internal coordinate is particle volume, this relation is readily written using volume balance as $x_r = x_1 + x_2$ where x_1 and x_2 are the volume of colliding particles and x_r is the volume of the resultant particle. In general, the internal coordinates of the two colliding particles and the resulting particle can be related by writing a suitable form of mass balance:

$$x_r = \zeta(x_1, x_2).$$

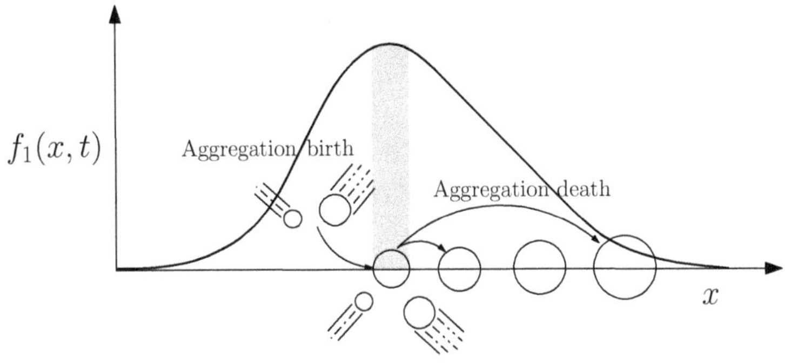

Figure 6.4 Birth and death of particles due to pure aggregation.

For example, if the particles are spherical liquid drops and the diameter is the internal coordinate, the function ζ is given by

$$x_r = \sqrt[3]{x_1^3 + x_2^3}.$$

A similar relation also exists for external coordinates:

$$\mathbf{r_r} = \rho(\mathbf{r}_1, \mathbf{r}_2).$$

The forms of this function can be obtained once the location of aggregate with respect to the colliding drops is defined. Let us consider a situation where gas bubbles are going up in a long vertical vessel. The external coordinate in this case is the height of the bubble from the base of the vessel, denoted by z. After aggregation, the aggregated bubble is located at the mean location of the two bubbles. Hence, the function ρ is given by

$$z_r = (z_1 + z_2)/2.$$

6.3.1 Number Balance for Aggregation

To develop the number balance for aggregation, we will consider the case where all other processes, such as nucleation, growth or breakage, are absent. Similar to previous cases, we will consider an elementary volume in the internal coordinate (dx) as well as in the external coordinate space, $d\mathbf{V_r}$. Figure 6.4 shows the situation where only the internal coordinate is shown. The number of particles in the elementary volume is $f_1(x, \mathbf{r}, t)dxd\mathbf{V_r}$ and the rate of change of the number of particles in this elementary strip is

$$\frac{df_1(x, \mathbf{r}, t)dxd\mathbf{V_r}}{dt}.$$

To write the balance equation, we need to account for aggregation death and birth terms in a similar fashion to what we did for breakage. The particles in this small strip around x will aggregate with other particles and count as a loss term. How many such losses (aggregation death) happen per unit time? That depends on the number of possible pairs and the efficiency of aggregation. The pairs that contribute to the loss term in the small zone of interest must have x and $\mathbf{r}$ as coordinates in one of the particles in the pair. The number of such pairs is given by $f_2(x, x', \mathbf{r}, \mathbf{r}', t)dxdx'd\mathbf{V}_{\mathbf{r}}d\mathbf{V}_{\mathbf{r}'}$, where x' and $\mathbf{r}'$ are arbitrary.

Not all such pairs undergo aggregation and the number of actual aggregations that occurs (i.e. the number of losses) per unit time is dependent on the frequency of aggregation. The number of actual aggregations therefore is given by

$$a(x, x', \mathbf{r}, \mathbf{r}', y(t))f_2(x, x', \mathbf{r}, \mathbf{r}', t)dxdx'd\mathbf{V}_{\mathbf{r}}d\mathbf{V}_{\mathbf{r}'}.$$

Because x' and $\mathbf{r}'$ are arbitrary, the total loss term, denoted by D_a (death due to aggregation), is found by integrating the above expression over the entire domain of x' and $\mathbf{r}'$:

$$D_a = \int_{\Omega_{x'}} \int_{\Omega_{r'}} a(x, x', \mathbf{r}, \mathbf{r}', y(t))f_2(x, x', \mathbf{r}, \mathbf{r}', t)dxdx'd\mathbf{V}_{\mathbf{r}}d\mathbf{V}_{\mathbf{r}'}.$$

There will also be aggregation birth (B_a) because smaller particles will aggregate and form particles in the size range of interest. Let us assume that particles at $\tilde{x}$ and $\tilde{\mathbf{r}}$ will aggregate with particles at x' and $\mathbf{r}'$ to produce particles at x and $\mathbf{r}$. For a given value of x and $\mathbf{r}$, $\tilde{x}$ and x' are related by the function ζ, and $\tilde{\mathbf{r}}$ and $\mathbf{r}'$ are related by the function ρ.

Hence, the number of possible aggregating pairs giving birth in the marked region would be

$$f_2(x', \tilde{x}, \mathbf{r}', \tilde{\mathbf{r}}, t)dx'd\tilde{x}d\mathbf{V}_{\mathbf{r}'}d\mathbf{V}_{\tilde{\mathbf{r}}}.$$

The number of aggregates formed at x and $\mathbf{r}$ is therefore obtained by using the aggregation frequency:

$$a(x', \tilde{x}, \mathbf{r}', \tilde{\mathbf{r}}, y(t))f_2(x', \tilde{x}, \mathbf{r}', \tilde{\mathbf{r}}, t)dx'd\tilde{x}d\mathbf{V}_{\mathbf{r}'}d\mathbf{V}_{\tilde{\mathbf{r}}}.$$

The total gain will therefore be given by

$$B_a = \frac{1}{2}\int_{\Omega_{x'}} \int_{\Omega_{r'}} a(x', \tilde{x}, \mathbf{r}', \tilde{\mathbf{r}}, y(t))f_2(x', \tilde{x}, \mathbf{r}', \tilde{\mathbf{r}}, t)dx'd\tilde{x}d\mathbf{V}_{\mathbf{r}'}d\mathbf{V}_{\tilde{\mathbf{r}}}.$$

The aggregation birth term counts the pairs and because of symmetry it counts each pair twice. The term $1/2$ is included to correct for this redundancy. The overall number balance for aggregation is therefore written as

$$\frac{df_1(x, \mathbf{r}, t)dxd\mathbf{V}_\mathbf{r}}{dt} =$$
$$\frac{1}{2}\int_{\Omega_{x'}}\int_{\Omega_{r'}} a(x', \tilde{x}, \mathbf{r}', \tilde{\mathbf{r}}, y(t))f_2(x', \tilde{x}, \mathbf{r}', \tilde{\mathbf{r}}, t)dx'd\tilde{x}d\mathbf{V}_{\mathbf{r}'}d\mathbf{V}_{\tilde{\mathbf{r}}}$$
$$-\int_{\Omega_{x'}}\int_{\Omega_{r'}} a(x, x', \mathbf{r}, \mathbf{r}', y(t))f_2(x, x', \mathbf{r}, \mathbf{r}', t)dxdx'd\mathbf{V}_\mathbf{r}d\mathbf{V}_{\mathbf{r}'}.$$

This is the population balance equation for aggregation, but this equation needs a series of simplifications before we can use it.

6.3.2 Simplification of the Aggregation Equation

The first step in simplifying this equation is to cancel the $dxd\mathbf{V}_\mathbf{r}$ terms. In the first integration term the differential is in terms of $\tilde{x}$ and $\tilde{\mathbf{r}}$ instead of x and $\mathbf{r}$, and hence a variable change from $\tilde{x}$ to x and $\tilde{\mathbf{r}}$ to $\mathbf{r}$ in the integration is required. This is possible because the functional relation between these variables is known through the functions ζ and ρ. The formal process is to write the Jacobian of the transformation[3] upon which the first integral term becomes

$$\frac{1}{2}\int_{\Omega_{x'}}\int_{\Omega_{r'}} a(x', \tilde{x}, \mathbf{r}', \tilde{\mathbf{r}}, y(t))f_2(x', \tilde{x}, \mathbf{r}', \tilde{\mathbf{r}}, t)dx', dx, d\mathbf{V}_{\mathbf{r}'}d\mathbf{V}_\mathbf{r}\left|\frac{\partial(\tilde{x}, \tilde{\mathbf{r}})}{\partial(x, \mathbf{r})}\right|.$$

The Jacobian can be expanded to

$$\frac{\partial(\tilde{\mathbf{x}}, \tilde{\mathbf{r}})}{\partial(\mathbf{x}, \mathbf{r})} = \begin{vmatrix} \frac{\partial\tilde{x}_1}{\partial x_1} & \cdots & \frac{\partial\tilde{x}_1}{\partial x_n} & \frac{\partial\tilde{x}_1}{\partial r_1} & \frac{\partial\tilde{x}_1}{\partial r_2} & \frac{\partial\tilde{x}_1}{\partial r_3} \\ \vdots & \vdots & \vdots & \vdots & \vdots & \vdots \\ \frac{\partial\tilde{x}_n}{\partial x_1} & \cdots & \frac{\partial\tilde{x}_n}{\partial x_n} & \frac{\partial\tilde{x}_n}{\partial r_1} & \frac{\partial\tilde{x}_n}{\partial r_2} & \frac{\partial\tilde{x}_n}{\partial r_3} \\ \frac{\partial\tilde{r}_1}{\partial x_1} & \cdots & \frac{\partial\tilde{r}_1}{\partial x_n} & \frac{\partial\tilde{r}_1}{\partial r_1} & \frac{\partial\tilde{r}_1}{\partial r_2} & \frac{\partial\tilde{r}_1}{\partial r_3} \\ \frac{\partial\tilde{r}_2}{\partial x_1} & \cdots & \frac{\partial\tilde{r}_2}{\partial x_n} & \frac{\partial\tilde{r}_2}{\partial r_1} & \frac{\partial\tilde{r}_2}{\partial r_2} & \frac{\partial\tilde{r}_2}{\partial r_3} \\ \frac{\partial\tilde{r}_3}{\partial x_1} & \cdots & \frac{\partial\tilde{r}_3}{\partial x_n} & \frac{\partial\tilde{r}_3}{\partial r_1} & \frac{\partial\tilde{r}_3}{\partial r_2} & \frac{\partial\tilde{r}_3}{\partial r_3} \end{vmatrix}.$$

Now the like differential terms can be cancelled and the equation becomes

$$\frac{\partial f_1(x, \mathbf{r}, t)}{\partial t} = \frac{1}{2}\int_{\Omega_{x'}}\int_{\Omega_{r'}} a(x', \tilde{x}, \mathbf{r}', \tilde{\mathbf{r}}, y(t))f_2(x', \tilde{x}, \mathbf{r}', \tilde{\mathbf{r}}, t)dx'd\mathbf{V}_{\mathbf{r}'}\left|\frac{\partial(\tilde{x}, \tilde{\mathbf{r}})}{\partial(x, \mathbf{r})}\right|$$
$$-\int_{\Omega_{x'}}\int_{\Omega_{r'}} a(x, x', \mathbf{r}, \mathbf{r}', y(t))f_2(x, x', \mathbf{r}, \mathbf{r}', t)dx', d\mathbf{V}_{\mathbf{r}'}.$$

The pair density function can be approximately written as the product of the singlet density:

$$f_2(x, x', \mathbf{r}, \mathbf{r}', t) \approx f_1(x, \mathbf{r}, t)f_1(x', \mathbf{r}', t).$$

3 See section A.4 for details of this transformation process.

Using this, the above equation becomes

$$\frac{\partial f_1(x,\mathbf{r},t)}{\partial t} = \frac{1}{2}\int_{\Omega_{x'}}\int_{\Omega_{r'}} a(x',\tilde{x},\mathbf{r}',\tilde{\mathbf{r}},y(t))f_1(\tilde{x},\tilde{\mathbf{r}},t)f_1(x',\mathbf{r}',t)dx'd\mathbf{V}_{\mathbf{r}'}\left|\frac{\partial(\tilde{x},\tilde{\mathbf{r}})}{\partial(x,\mathbf{r})}\right|$$
$$-\int_{\Omega_{x'}}\int_{\Omega_{r'}} a(x,x',\mathbf{r},\mathbf{r}',y(t))f_1(x,\mathbf{r},t)f_1(x',\mathbf{r}',t)dx',d\mathbf{V}_{\mathbf{r}'}.$$

Further, if the system is well mixed, the number densities are not a function of the space variable. Hence, their spatial dependance may be dropped. With this the Jacobian also becomes much simpler, but the aggregation frequency inherently is a function of the spacial locations of the particles and spatial dependance cannot be dropped from the aggregation frequency:

$$\frac{\partial f_1(x,t)}{\partial t} = \frac{1}{2}\int_{\Omega_{x'}}\int_{\Omega_{r'}} a(x',\tilde{x},\mathbf{r}',\tilde{\mathbf{r}},y(t))f_1(\tilde{x},t)f_1(x',t)dx'd\mathbf{V}_{\mathbf{r}'}\left|\frac{\partial\tilde{x}}{\partial x}\right|$$
$$-\int_{\Omega_{x'}}\int_{\Omega_{r'}} a(x,x',\mathbf{r},\mathbf{r}',y(t))f_1(x,t)f_1(x',t)dx',d\mathbf{V}_{\mathbf{r}'}.$$

Now, multiplying both sides of the above equation by $d\mathbf{V}_\mathbf{r}$ and integrating wrt $\mathbf{r}$ we have

$$V(\Omega_\mathbf{r})\frac{\partial f_1(x,t)}{\partial t} = \frac{1}{2}\int_{\Omega_{x'}}\int_{\Omega_{r}}\int_{\Omega_{r'}} a(x',\tilde{x},\mathbf{r}',\tilde{\mathbf{r}},y(t))f_1(\tilde{x},t)f_1(x',t)dx'd\mathbf{V}_{\mathbf{r}'}d\mathbf{V}_{\mathbf{r}}\left|\frac{\partial\tilde{x}}{\partial x}\right|$$
$$-\int_{\Omega_{x'}}\int_{\Omega_{r}}\int_{\Omega_{r'}} a(x,x',\mathbf{r},\mathbf{r}',y(t))f_1(x,t)f_1(x',t)dx',d\mathbf{V}_{\mathbf{r}'}d\mathbf{V}_{\mathbf{r}}.$$

Defining

$$a(x,x') = \frac{\int_{\Omega_r}\int_{\Omega_{r'}} a(x,x',\mathbf{r},\mathbf{r}',y(t))d\mathbf{V}_{\mathbf{r}'}d\mathbf{V}_{\mathbf{r}}}{V(\Omega_\mathbf{r})}$$

the above equation becomes

$$\frac{\partial f_1(x,t)}{\partial t} = \frac{1}{2}\int_{\Omega_{x'}} a(\tilde{x},x',y(t))f_1(\tilde{x},t)f_1(x',t)dx'\left|\frac{\partial\tilde{x}}{\partial x}\right|$$
$$-\int_{\Omega_{x'}} a(x,x',y(t))f_1(x,t)f_1(x',t)dx'.$$

Because the external coordinates are not important, only the $d\tilde{x}/dx$ term in the Jacobian exists. If the internal coordinate is volume, the function ζ becomes

$$\tilde{x} = x - x'$$

and hence the Jacobian is unity. If we also assume that the continuous phase does not affect the aggregation, we get the most simplified form of the aggregation equation:

$$\frac{\partial f_1(x,t)}{\partial t} = \frac{1}{2}\int_{x'=0}^{x'=x} a(x-x',x')f_1(x-x',t)f_1(x',t)dx' - \int_{x'=0}^{x'=\infty} a(x,x')f_1(x,t)f_1(x',t)dx'. \tag{6.7}$$

Note that limits have been explicitly written now. It should be remembered that the above equation is not a general case and a series of simplifications has been used to arrive at this form.

Example 6.5 Consider a very long, thin, vertical pipe filled with water. At the bottom of the vessel, a sparger is releasing nitrogen. Dissolved oxygen from the water enters the nitrogen bubbles and reduces the oxygen level in the water. A temperature gradient exists in the tube and hence the external coordinate will be important. The bubbles also aggregate while rising in the long tube. Breakage is not observed. Define a suitable internal coordinate vector for this case and find out the Jacobian to be used in the aggregation population balance equation.

Solution: The nitrogen bubbles are the dispersed phase, with the volume of the bubble (x) and the concentration of oxygen in it (c) as two internal coordinates. Since the tube is thin and long, the elevation from the bottom of the tube (z) can be taken as the only external coordinate. Next, we have to find out the required relations between the internal and external coordinates of the aggregating drops and the resulting drop. We shall consider binary collisions only. If the density is relatively constant, the overall mass balance can be written as

$$\tilde{x} + x' = x,$$

which gives

$$\tilde{x} = x - x'.$$

Writing the component mass balance for oxygen in the bubble gives

$$\tilde{x}\tilde{c} + x'c' = xc.$$

Substituting for $\tilde{x}$ and rearranging,

$$\tilde{c} = \frac{xc - x'c'}{x - x'}.$$

If the location of the newly formed bubble is at the mean location of the two aggregating bubbles, the corresponding relation for z is given by

$$\tilde{z} = 2z - z'.$$

Hence, the Jacobian is given by

$$\begin{vmatrix} \frac{\partial \tilde{x}}{\partial x} & \frac{\partial \tilde{x}}{\partial c} & \frac{\partial \tilde{x}}{\partial z} \\ \frac{\partial \tilde{c}}{\partial x} & \frac{\partial \tilde{c}}{\partial c} & \frac{\partial \tilde{c}}{\partial z} \\ \frac{\partial \tilde{z}}{\partial x} & \frac{\partial \tilde{z}}{\partial c} & \frac{\partial \tilde{z}}{\partial z} \end{vmatrix} = \begin{vmatrix} 1 & 0 & 0 \\ \frac{x'(c'-c)}{(x-x')^2} & \frac{x}{x-x'} & 0 \\ 0 & 0 & 2 \end{vmatrix} = \frac{2x}{x-x'}.$$

□

6.3.3 Models for Aggregation Frequency

Estimating the frequency of aggregation is a complex task. Fortunately, aggregation frequencies for a large number of processes can be found in the literature. If the aggregation frequency of the desired process is not available, detailed phenomenological modeling has to be carried out. An example of such a process was demonstrated in section 4.5.2 where we derived the aggregation frequency for the purely diffusive motion of mono sized colloidal particles. If the particles are polydispersed, the aggregation frequency for a pair is given by the well-known Brownian kernel:

$$a(x,x') = \frac{2kT}{3\mu}(x'^{1/3} + x^{1/3})(x'^{-1/3} + x^{-1/3}).$$

Another example of aggregation frequency is for gravity settling of particles, which is given by [3]

$$a(x,x') = \frac{2(\rho_p - \rho)g}{9\mu}(3\sqrt{\pi}/4)^{2/3}(x'^{1/3} + x^{1/3})|x'^2 - x^2|.$$

In both cases, particle volume is taken as internal co-ordinate. It can be observed that if the volumes of the particles are same, they settle at the same velocity and hence the frequency of aggregation for such pairs is zero. In absence of any mechanistic kernels, empirical kernels such as constant ($a(x,x') = a_0$), sum ($a(x,x') = a_0(x,+x')$) and product ($a(x,x') = a_0 xx'$) kernels are used. These three kernels are arranged in increasing order of particle size dependence.

Example 6.6 [4]The name 'population balance' suggests that this tool can be used for analysis of demographic information. One of the important information that demographers ask for is the age distribution in a human population. Depending on this information, many socioeconomic decisions are made. Formulate a population balance model to predict the steady state population distribution considering age as internal coordinate. What fraction of the population is expected to live to at least 70 years?

---

4 Adopted from [4].

Data: Government records show that the death rate (as a fraction of the population) as a function of age (τ) may be written as

$$D(\tau) = 0.005\left[1 + 2\left(\frac{\tau}{40}\right)^4\right]$$

and the birth propensity function $\rho(\tau)$ is found to be a log-normal probability density function

$$\rho(\tau) = \frac{1}{\sqrt{2\pi}\log\sigma'}\exp\left[-\frac{\log^2(\tau/25)}{2\log^2\sigma'}\right],$$

which peaks at age 25. The spread parameter σ' is usually found to be 1.2. The population is isolated, i.e. no immigration or emigration is permitted.

The number of births from a population of age τ is the product of the fraction of people living up to age τ and the propensity of them to give birth. Hence,

$$B_{tot} = \int_0^\infty B^0(\tau)d\tau = k\int_0^\infty \frac{f_1(\tau,t)}{f_1(0,t-\tau)}\rho(\tau)d\tau.$$

Here k is the overall fertility factor, which depends on the amount of resources available.

Solution: It can be readily seen that only one attribute of the population is of interest: age. Hence, only one internal coordinate, age, should be considered. Birth, death, and growth are the three processes that can change the age distribution of the population. Of these processes, birth can be treated as a boundary condition since it resembles the nucleation process. Just as nuclei are born with zero size, people are born with zero age. Hence, the population balance equation can be written as

$$\frac{\partial f_1(\tau,t)}{\partial t} + v_\tau\frac{\partial f_1(\tau,t)}{\partial \tau} + D(\tau)f_1(\tau,t) = 0.$$

Here the function $D(\tau)$ has similar significance to the breakage frequency. The initial distribution $f_1(\tau,0)$ is known and the boundary condition becomes

$$B_{tot} = v_0 f_1(0,t).$$

Recognizing that $v_\tau = 1$, the steady state balance equation becomes

$$\frac{df_{1s}(\tau)}{d\tau} + D(\tau)f_{1s}(\tau) = 0,$$

where f_{1s} is the steady state number density. The boundary condition can be written for the steady state as

$$f_{1s}(0) = k\int_0^\infty \frac{f_{1s}(\tau)}{f_{1s}(0)}\rho(\tau)d\tau.$$

Now we can solve the population balance equation to obtain the steady state number density function:

$$\frac{f_{1s}(\tau)}{f_{1s}(0)} = \exp\left[-\int_0^{\tau} D(\tau')d\tau'\right].$$

In this expression, $f_{1s}(0)$ is the unknown constant, which can be obtained by using the boundary condition

$$f_{1s}(0) = k\int_0^{\infty} \exp\left[-\int_0^{\theta} D(\tau')d\tau'\right]\rho(\tau)d\tau.$$

The steady state fraction of people who are alive at age 70 is given by

$$\frac{f_{1s}(70)}{f_{1s}(0)} = \exp\left[-\int_0^{70} D(\tau)d\tau\right] = 0.189.$$

□

6.4 Exercises

Exercises 6.1 Shear-induced degradation of polymers is an important process. Polymer chains tend to break due to the shear field in the processing equipment and the molecular weight of the polymer decreases. One such situation where breakage happens due to flow in a capillary is shown in Figure 6.5.
The degradation reaction can be written as:

$$P(x) \rightarrow P(x/2).$$

The rate constant for this degradation reaction is a function of molecular weight and is given by

$$k(x) = \kappa(x - x_l),$$

where κ is independent of molecular weight and x_l is the limiting molecular weight below which degradation does not occur. Write down the population balance model for a batch process involving only the shear-induced degradation of polymers.

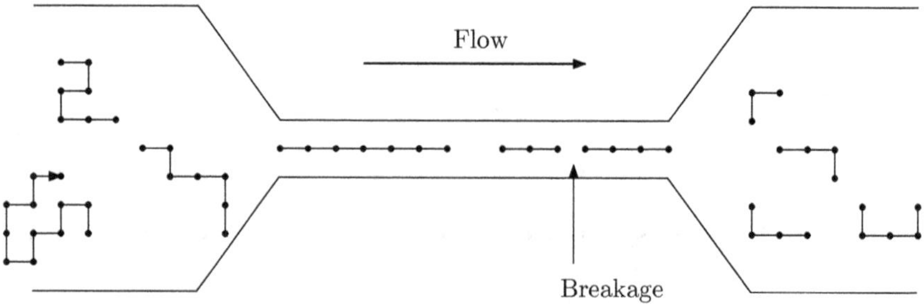

Figure 6.5 Extensional degradation of polymers.

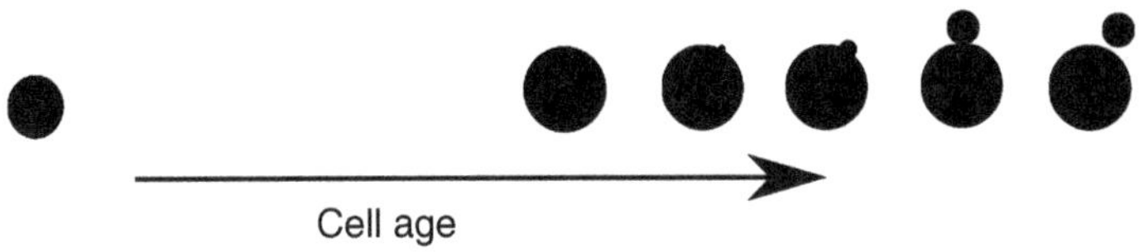

Figure 6.6 Cell cycle.

Exercises 6.2 The cells of a particular bacterial species grow and divide as shown in Figure 6.6. The newly born daughter cells have a distribution of sizes. The cell culture grows in a batch fermentor containing a nutrient and a product is formed *inside* the cells. Cell growth is dependent on cell mass and nutrient concentration. Formulate a population balance model to predict the transient cell mass distribution and nutrient concentration.

An improved version of this process is also used where a drug is added along with the nutrient. In the absence of the drug, the product is stored inside the cells and is difficult to extract. The drug affects the cell membrane in such a way that the product leaks outside the cells. Since this reduces the product concentration inside the cells, more product is formed by the cells, but the drug also kills some of the otherwise healthy cells. However, the growth rate remains unaffected by the drug. Modify the previous model so that it can be used for the improved process.

Exercises 6.3 Consider a fermenter (well mixed) where microbial population is growing in a nutrient solution. The fermentor is fed with fresh cells and a nutrient, and an outlet stream is drawn from the fermenter. Cell division, growth, and death occur in the fermenter. The dead cells decompose and finally produce nutrients for other cells. The typical time required for complete degradation of a dead cell is τ_d. The nutrient produced by dead cells may be assumed to be identical to that supplied. The growth and death rates of the cells are dependent on cell mass and nutrient concentration. Formulate a population balance model to predict the steady state concentration of the cells (in terms of cell mass) in the outlet stream.

Exercises 6.4 A mixture of water and oil is stirred vigorously to produce an emulsion. Under certain conditions, a steady state drop size distribution is achieved. Write down the population balance equation to predict the steady state drop size distribution assuming the drops are perfect spheres. Use drop volume as the internal coordinate. Will the population balance equation be different if the diameter of the drops is taken as the internal coordinate instead of volume? Explain.

Exercises 6.5 Imagine yourself as an R&D engineer in a food product manufacturing plant. Your plant produces a cereal called 'Granola Grin', which is

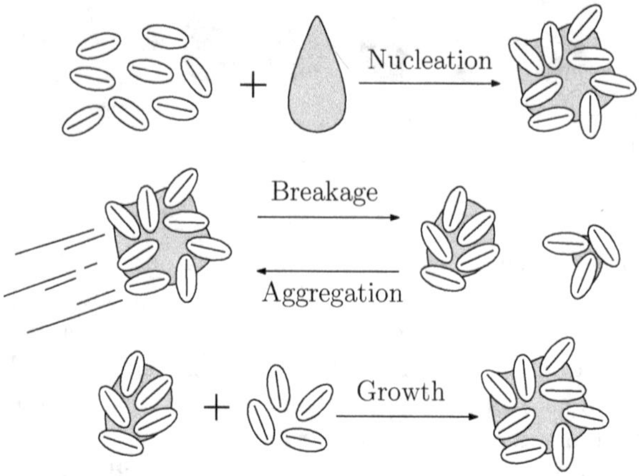

Figure 6.7 Making 'Granola Grin'.

sweetened granola nuggets. Roasted granola comes from a roasting unit and undergoes a process called granulation in the plant under your supervision. The process is a batch process. First, the roasted granola (approximately 5 mm in size) is fed into a rotating drum. Soon after, a rotating bucket sprays the roasted granola with a small amount of hot sugar syrup for 2 minutes. The sugar syrup makes small aggregates of size 2 cm. The drum continues to rotate slowly at 15 rpm for another 10 minutes then the rotational speed is reduced to 5 rpm and hot air is circulated through the drum for another 5 minutes. Finally, the product goes to the packaging section.

Despite the simple nature of this process, the manager has noted that it is very difficult to get rid of the 'dust', which is formed from single granola seeds coated with sugar. The 'dust' is difficult to recycle as it produces a much sweeter product. Because of this your company is loosing market share to its competitor, 'Granny's Granola'. The company has tried various things to reduce the amount of 'dust' produced, such as changing the rotational speed or air flow rate etc., but with no success. They want to give modeling a try and ask you to formulate a population balance model for the process. When you interview the shift supervisor, he says, 'I can solve the problem if I'm allowed to spray more sugar syrup but the quality control department won't allow it as it spoils the crispy nature of the nuggets.' The process for making 'Granola Grin' may contain one or more of the steps shown in Figure 6.7.

1) What are the internal variables that are needed to represent the particles?
2) Are external coordinates important?
3) Are all these processes important or do some of them dominate? Justify your answer.

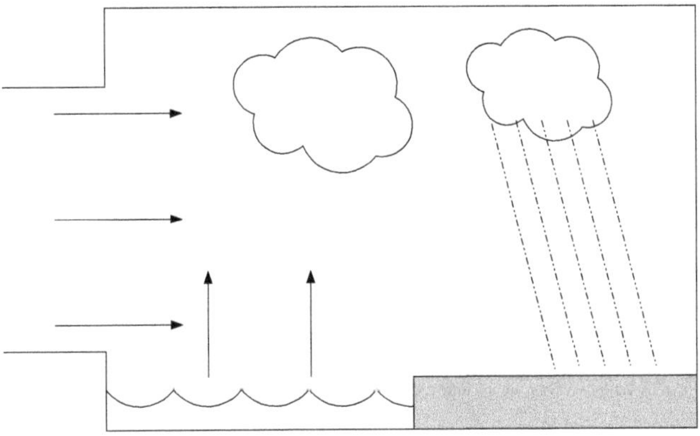

Figure 6.8 A controlled environment chamber.

4) Develop a population balance model for the above process. Make all necessary assumptions and justify them.
5) Suggest some way to minimize the fines on the basis of your model.

Exercises 6.6 Reactive precipitation is a process where two chemicals A and B, soluble in the solvent S, react to give a sparingly soluble product C. The product nucleates and grows to produce matured particles. The small particles produced during nucleation or the early stage of growth may form aggregates. To prevent this, a stabilizer is usually added prior to the reaction. The stabilizer is adsorbed on the particle surface so that part of the surface is covered by stabilizer and becomes unavailable for growth. This coverage also prevents aggregation. Very small particles are formed in reactive precipitation and breakage is unlikely. Model this system to predict the final particle size distribution of a batch process.

Exercises 6.7 In the controlled environment chamber shown in Figure 6.8 water vapour is formed at the lower part and cooled at the upper part to a temperature sufficiently below its dew point. Small drops condense on dust particles in the upper portion and grow by both further condensation and aggregation. Once the drops become larger than 1 μm, they start to fall by gravity. Note that the drops form a stable aerosol in the upper portion of the chamber.

1) Determine an appropriate set of internal and external coordinates and a continuous phase vector for this system.
2) Write down a population balance model.
3) Search the literature to investigate the mechanism of coalescence and/or breakage of such drops.

4) Comment on the form of (i) the breakage functions, (ii) the aggregation frequency, and (iii) the growth rates to be used.

Exercises 6.8 Bread dough is a very viscous medium which contains air pockets in a continuous phase of starch particles in water. The air pockets are introduced during the mixing of wheat flour and water. After formation, the dough is proved. During proving, yeast cells produce CO_2, which makes the mixture more porous. There is an argument going on at PBM Bakery about whether mixing or proving is the most important process in determining bread quality. To resolve this issue, food process engineers at PBM Bakery need your help in writing a population balance model. They provide the following information:

- The dough is very viscous and bubble movement is virtually impossible. However, two nearby bubbles can grow and coalesce.
- During fermentation no new bubbles are created, only the existing ones grow.

It is also known that bread porosity is affected by the production and transport of CO_2.

Formulate a population balance model to predict the bubble size distribution as a function of proving time.

Bibliography

1 R. P. Chhabra and G. Basavaraj. *Coulson and Richardson's Chemical Engineering: Particulate Systems and Particle Technology*, volume 2A, 6th edition. Elsevier, 2019.

2 L. Vogel and W. Peukert. Breakage behaviour of different materials: Construction of a mastercurve for the breakage probability. *Powder Technology*, *129*(1-3):101–110, 2003.

3 D. Ramkrishna. *Population Balances: Theory and applications to particulate systems in engineering*. Academic Press, 2000.

4 A. D. Randolph and M. A. Larson. *Theory of Particulate Processes: Analysis and Techniques of Continuous Crystallization*, 2nd edition. Academic Press, 1988.

7

Solution of the Population Balance Equation

The population balance equation (PBE) formulated in the previous chapters is a complicated equation. For the growth process it is a partial differential equation (PDE), for breakage it is a partial integro-differential equation, and for aggregation it is a non-linear partial integro-differential equation. We have also seen that it is often coupled with other models such as the growth rate of individual crystals, breakage rate and daughter distribution functions, aggregation frequency or mass balance involving continuous phase vector. Together they form a very complicated set and numerical solution is inevitable. However, analytical solution for a few simplified cases can be obtained and such solutions are of great importance because they are needed for benchmarking the numerical solution.

Another important aspect of the PBE is the equations of the moments obtained from it. Moments are integrals of the weighted particle size distribution (PSD) and are related to various measurable properties. Moments are also used for benchmarking the numerical solution.

In this chapter we shall first discuss the technique for obtaining equations for moments of the PBE and then discuss a few analytical solution techniques. Towards the end of the chapter a popular numerical technique for solving the PBE will be discussed.

7.1 Operations Involving Moments of the PBE

We have already seen (section 5.3.1) that the integration of the number density over the entire range of internal state variables yields the total number of particles:

$$N(t) = \int_0^\infty f_1(x,t)dx.$$

Engineering of Submicron Particles: Fundamental Concepts and Models, First Edition. Jayanta Chakraborty.

Companion website: http://booksupport.wiley.com

Similarly, we can also obtain the total mass of particles by multiplying the particle mass with the number density and integrating over the complete internal coordinate space. If the internal coordinate is particle mass:

$$M(t) = \int_0^\infty x f_1(x,t)dx.$$

These are called moments. In general, the jth moment is defined as

$$\mu_j(t) \equiv \int_0^\infty x^j f_1(x,t)dx.$$

Apart from total number (μ_0) and mass (μ_1), many other useful properties of the PSD, such as average mass (μ_1/μ_0), variance or skewness, can be obtained from various moments. Even the PSD can be reconstructed approximately if a number of moments are known [1].

It is clear from the definition that moments are a function of time and the equation describing the evolution of moments can be obtained from the PBE itself. Let us consider the breakage PBE and see how the equation for the jth moment can be obtained:

$$\frac{\partial f_1(x,t)}{\partial t} = -\Gamma(x) f_1(x,t) + \int_x^\infty \Gamma(x') f_1(x',t)\nu(x')P(x|x')dx'.$$

Here the internal coordinate x is particle mass. Let us consider the following breakage functions:

$$\begin{aligned} \nu(x) &= 2.0 \\ \Gamma(x) &= kx \\ P(x|x') &= 1/x'. \end{aligned}$$

Multiplying both sides of this equation by x^j and integrating from 0 to ∞,

$$\begin{aligned} \frac{d\mu_j(t)}{dt} &= -\int_0^\infty kx^{j+1} f_1(x,t) + 2\int_0^\infty \int_x^\infty x^j kx' f_1(x',t)\frac{1}{x'}dxdx' \\ &= -k\mu_{j+1}(t) + 2k\int_{x=0}^\infty \int_{x'=x}^\infty x^j f_1(x',t)dxdx'. \end{aligned}$$

The number density in the second integral is in x' but the limit for this variable is not from 0 to ∞ and hence it cannot be written in terms of a moment. The standard remedy to this problem is to change the limits of the integral, i.e. cover the same domain in a different order, as shown in Figure 7.1.

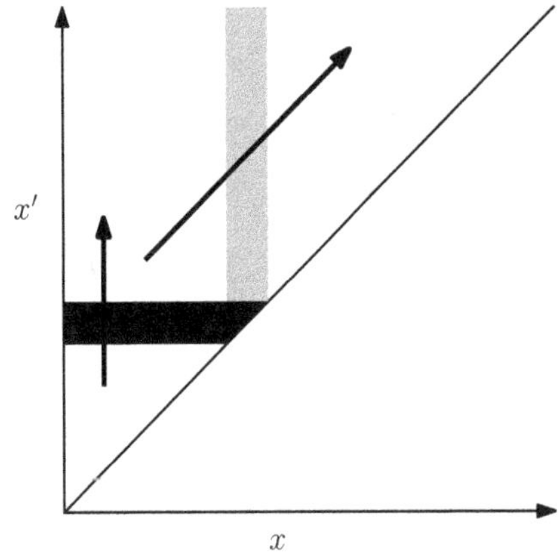

Figure 7.1 Change of limit for moment operation. Gray indicates the original order in which the domain is covered and black indicates the new order. The line in the middle of the domain represents $x = x'$.

With this, the equation becomes

$$\begin{aligned}
\frac{d\mu_j(t)}{dt} &= -k\mu_{j+1}(t) + 2k\int_{x'=0}^{\infty}\int_{x=0}^{x'} f_1(x',t)x^j dx dx' \\
&= -k\mu_{j+1}(t) + 2k\int_0^{\infty} f_1(x',t)\left[\int_0^{x'} x^j dx\right] dx' \\
&= -k\mu_{j+1}(t) + \frac{2k}{j+1}\int_0^{\infty} x'^{j+1} f_1(x',t)dx' \\
&= -k\mu_{j+1}(t) + \frac{2k}{j+1}\mu_{j+1}(t) \\
&= -k\frac{j-1}{j+1}\mu_{j+1}(t).
\end{aligned}$$

It can be seen that the equation for the jth moment is dependent on the $(j+1)$th moment. This is usually the case and no closed-form solution is possible for the moments. This is called the closure problem. However, for this case, $\mu_1(t)$ is the total mass of particles and must be a constant. Hence the evolution of the zeroth moment can be found. It is clear from the above equation that the zeroth moment varies linearly with time.

Example 7.1 Rework the above problem for the mid-point breakage. All other breakage functions remain the same.

Solution: For mid-point breakage, the daughter distribution function is given by

$$P(x|x') = \delta\left(x - \frac{x'}{2}\right).$$

The moment of the breakage equation becomes

$$\frac{d\mu_j(t)}{dt} = -\int_0^\infty kx^{j+1}f_1(x,t) + 2\int_0^\infty \int_x^\infty x^j kx' f_1(x',t)\delta\left(x - \frac{x'}{2}\right) dxdx'.$$

Changing the limit as before:

$$\begin{aligned}\frac{d\mu_j(t)}{dt} &= -k\mu_{j+1}(t) + 2k\int_0^\infty dx' \int_0^{x'} dxx'x^j\delta\left(x - \frac{x'}{2}\right) f_1(x',t)\\ &= -k\mu_{j+1}(t) + 2k\int_0^\infty dx'x'\left(\frac{x'}{2}\right)^j f_1(x',t)\\ &= -k\mu_{j+1}(t) + 2^{1-j}k\mu_{j+1}(t).\end{aligned}$$

As expected, μ_1 is constant. □

7.2 Analytical Solutions of the PBE

An analytical solution of the PBE is possible for some cases. The most common way to obtain an analytical solution is to assume that only one of the processes is present. A range of methods have been used for the analytical solution of the PBE and often one method is best for a particular kind of equation. For example, the pure growth equation may be solved readily using the method of characteristics. On the other hand, the aggregation equation may be solved using the method of Laplace transform because of the presence of the non-linear convolution term. Breakage kernels often show similarity behavior and similarity solution is a popular method for analytical solution of breakage equations. We shall discuss these techniques in turn now.

7.2.1 Solution of the Growth Equation: Method of Characteristics

Let us consider the following PBE involving a size-dependent growth process:[1]

$$\frac{\partial f_1}{\partial t} + \frac{\partial[-\alpha x^{\frac{2}{3}} f_1]}{\partial x} = 0.$$

First let us reduce this equation to the standard form

$$a(x,t)f_x(x,t) + b(x,t)f_t(x,t) = c(x,t)$$

by substituting $x^{\frac{2}{3}} f_1 = f$. Here the subscript to the function indicates the derivative. With this substitution, this equation becomes

$$x^{-\frac{2}{3}}\frac{\partial f}{\partial t} + (-\alpha)\frac{\partial f}{\partial x} = 0.$$

1 For details of the method of characteristics, see section A.5.

The auxiliary equation for this PDE is

$$\frac{dt}{x^{-\frac{2}{3}}} = \frac{dx}{-\alpha} = \frac{df}{0}.$$

From the first pair,

$$x^{\frac{1}{3}} + \frac{\alpha}{3}t = c.$$

From the second pair,

$$f = \phi(c) = \phi\left(x^{\frac{1}{3}} + \frac{\alpha}{3}t\right).$$

Transforming back:

$$f_1 = x^{-\frac{2}{3}}\phi\left(x^{\frac{1}{3}} + \frac{\alpha}{3}t\right).$$

Satisfying the initial condition:

$$f_{10}(x) = x^{-\frac{2}{3}}\phi(x^{\frac{1}{3}})$$

$$\phi(x^{\frac{1}{3}}) = x^{\frac{2}{3}}f_{10}(x)$$

$$\phi(\zeta) = \zeta^2 f_{10}(\zeta^3)$$

$$f_1 = x^{-\frac{2}{3}}\left(x^{\frac{1}{3}} + \frac{\alpha}{3}t\right)^2 f_{10}\left[\left(x^{\frac{1}{3}} + \frac{\alpha}{3}t\right)^3\right].$$

7.2.2 Solution of the Aggregation Equation: Method of Laplace Transforms

Let us consider the aggregation equation with constant kernel:

$$\frac{\partial f_1(x,t)}{\partial t} = \frac{1}{2}\int_{x'=0}^{x'=x} a_0 f_1(x-x',t)f_1(x',t)dx' - \int_{x'=0}^{x'=\infty} a_0 f_1(x,t)f_1(x',t)dx'$$

with the initial condition as

$$f_1(x,0) = N_0 g(x),$$

where N_0 is the initial total number of particles and $g(x)$ is a normalized distribution for particle size.

Now, taking the Laplace transform of this equation and defining

$$\bar{f}(s,t) \equiv \int_0^\infty e^{-sx} f_1(x,t)dx$$

we get

$$\frac{\partial \bar{f}(s,t)}{\partial t} = \frac{a_0}{2}\int_0^\infty e^{-sx}dx \int_0^x f_1(x-x',t)f_1(x',t)dx'$$
$$- a_0 \int_0^\infty e^{-sx}dx f_1(x,t)\int_0^\infty f_1(x',t)dx'.$$

Changing the limit of the first integral term,

$$\frac{\partial \bar{f}(s,t)}{\partial t} = \frac{a_0}{2}\int_0^\infty dx' \int_{x=x'}^\infty dx e^{-sx} f_1(x-x',t) f_1(x',t) - a_0 \int_0^\infty e^{-sx} dx f_1(x,t) \int_0^\infty f_1(x',t)dx'.$$

Substituting $x - x' = u$,

$$\frac{\partial \bar{f}(s,t)}{\partial t} = \frac{a_0}{2}\int_0^\infty e^{-sx'} f_1(x',t)dx' \int_0^\infty e^{-su} f_1(u,t)du - a_0 \int_0^\infty e^{-sx} dx f_1(x,t) \int_0^\infty f_1(x',t)dx'.$$

Finally,

$$\frac{\partial \bar{f}(s,t)}{\partial t} = \frac{a_0}{2}\bar{f}(x,t)^2 - a_0 N(t)\bar{f}(x,t). \tag{7.1}$$

The required initial condition for this equation can be found by taking the Laplace transform of the initial condition:

$$\bar{f}(s,0) = N_0 \bar{g}(s).$$

Because $N(t)$ is the total number of particles, its evolution can be obtained from the equation of the zeroth moment of the aggregation equation and is given by

$$\frac{dN(t)}{dt} = -\frac{a_0}{2}N(t)^2, \qquad N(0) = N_0,$$

whose solution is

$$N(t) = \frac{N_o}{1 - 0.5 a_0 N_0 t}.$$

The non-linear ordinary differential equation (ODE) (eqn 7.1) can be transformed to the linear ODE by writing it in terms of $1/\bar{f}(s,t)$:

$$\frac{\partial [1/\bar{f}(s,t)]}{\partial t} = -\frac{a_0}{2} + a_0 N(t)([1/\bar{f}(s,t)]).$$

Although this is a PDE, no derivative exists in s and hence it can be solved using an integrating factor. Once $\bar{f}(s,t)$ is obtained, $f(x,t)$ can be obtained by inverting it. This portion is left as an exercise for the reader.

7.2.3 Solution of the Breakage Equation: Similarity Solution

Let us consider the breakage equation

$$\frac{\partial f_1(x,t)}{\partial t} = -\Gamma(x) f_1(x,t) + \int_x^\infty \Gamma(x') f_1(x',t) \nu(x') P(x|x')dx', \tag{7.2}$$

where the breakage functions are given by

$$\begin{aligned} \nu(x) &= 2.0 \\ \Gamma(x) &= kx \\ P(x|x') &= 1/x'. \end{aligned}$$

We will solve this equation using the method of combination of variables, which is also known as the similarity solution or the scaling solution. The reader might be familiar with this method from the solution of the transient transport equation. One important limitation of the similarity solution must be noted at the beginning: the similarity solution works for certain forms of breakage kernels only. For details on the similarity behaviour of the PBE the reader is referred to the text by Ramkrishna [2].

A monotonic function is more suitable for similarity analysis and hence the first step for solution is to transform the breakage equation in terms of a cumulative of the PSD. The cumulative function that will be used is the mass fraction undersize defined as

$$F(x,t) = \frac{\int_0^x x' f_1(x',t)dx'}{\int_0^\infty x' f_1(x',t)dx'} = \frac{\int_0^x x' f_1(x',t)dx'}{\mu_1}.$$

Again, we can consider the internal coordinate to be particle mass and hence the first moment is constant. Our first task will be to obtain the breakage number balance equation in terms of $F(x,t)$.

7.2.3.1 Breakage Equation in Terms of Mass Fraction Undersize

To obtain the equation for $F(x,t)$, we first change the variable x to ζ in eqn 7.2, multiply both sides by $\zeta d\zeta$, and integrate from 0 to x:

$$\begin{aligned} \int_0^x \zeta d\zeta \frac{\partial f_1(\zeta,t)}{\partial t} &= -\int_0^x \zeta d\zeta \Gamma(\zeta) f_1(\zeta,t) \\ &\quad + \int_{\zeta=0}^x \zeta d\zeta \int_{x'=\zeta}^\infty \Gamma(x') f_1(x',t) \nu(x') P(\zeta|x') dx', \end{aligned}$$

which becomes

$$\begin{aligned} \mu_1 \frac{\partial F(x,t)}{\partial t} &= -\int_0^x \zeta \Gamma(\zeta) f_1(\zeta,t) d\zeta \\ &\quad + \int_{\zeta=0}^x \int_{x'=\zeta}^\infty \Gamma(x') f_1(x',t) \nu(x') P(\zeta|x') \zeta d\zeta dx'. \end{aligned}$$

To express the right-hand side in terms of $F(x,t)$, we need a limit of 0 to x on x'. Hence we must change the limits of the second integration. It is clear from Figure 7.2 that the change in limits will involve one finite and one infinite domain. Dividing the integral into these two domains and simultaneously changing the limits:

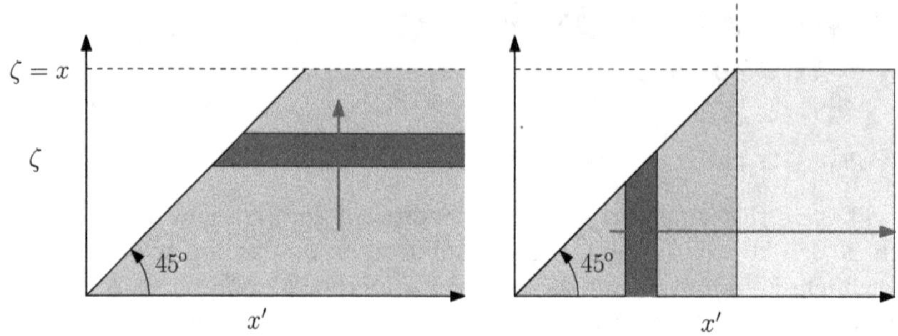

Figure 7.2 Division of the domain into one finite and one infinite domain, and change in order of integration. The 45° line represents $x' = \zeta$.

$$\mu_1 \frac{\partial F(x,t)}{\partial t} = -\int_0^x \zeta\Gamma(\zeta)f_1(\zeta,t)d\zeta + \int_{\zeta=0}^{x'}\int_{x'=0}^{x}\Gamma(x')f_1(x',t)\nu(x')P(\zeta|x')\zeta d\zeta dx' + \int_{\zeta=0}^{x}\int_{x'=x}^{\infty}\Gamma(x')f_1(x',t)\nu(x')P(\zeta|x')\zeta d\zeta dx'.$$

Rearranging:

$$\mu_1 \frac{\partial F(x,t)}{\partial t} = -\int_0^x \zeta\Gamma(\zeta)f_1(\zeta,t)d\zeta + \int_{x'=0}^{x}\Gamma(x')f_1(x',t)\left[\nu(x')\int_{\zeta=0}^{x'}P(\zeta|x')\zeta d\zeta\right]dx' + \int_{\zeta=0}^{x}\int_{x'=x}^{\infty}\Gamma(x')f_1(x',t)\nu(x')P(\zeta|x')\zeta d\zeta dx'.$$

Using the mass conservation equation (eqn 6.3) for breakage, the term in the square bracket becomes x' and hence the second term cancels with the first term and the equation becomes

$$\mu_1 \frac{\partial F(x,t)}{\partial t} = \int_{\zeta=0}^{x}\int_{x'=x}^{\infty}\Gamma(x')f_1(x',t)\nu(x')P(\zeta|x')\zeta d\zeta dx',$$

which can be written as

$$\frac{\partial F(x,t)}{\partial t} = \int_{x'=x}^{\infty}\Gamma(x')\frac{x'f_1(x',t)}{\mu_1}\left[\int_{\zeta=0}^{x}\frac{\nu(x')}{x'}P(\zeta|x')\zeta d\zeta\right]dx'$$

using the defining equation for $F(x,t)$,

$$\frac{\partial F(x,t)}{\partial t} = \int_x^{\infty}\Gamma(x')G(x|x')\frac{\partial F_1(x',t)}{\partial x'}dx', \tag{7.3}$$

where

$$G(x|x') = \frac{\nu(x')}{x'}\int_0^x d\zeta\zeta P(\zeta|x').$$

This completes our task of expressing the number balance in terms of a monotonic function.

7.2.3.2 Self Similar Form of the Breakage Equation

Our next task is to define a suitable variable (similarity variable) combining x and t so that the PDE (eqn 7.3) reduces to an ordinary differential equation. It is not possible to provide the form of the similarity variable a priori. It is customary to assume a generalized form and proceed by substituting the form into the equation and examining the situation. Hence, we shall assume a self similar solution $F(x,t) = \phi(\eta)$. The similarity variable η is given by a general form $\eta = x/h(t)$ where the function $h(t)$ is not known yet.

At this stage we shall restrict the solution to the particular breakage functions mentioned at the beginning of this section. With these breakage functions it can be seen that

$$G(x|x') = \left(\frac{x}{x'}\right)^2$$

and the breakage PBE becomes

$$\frac{\partial F(x,t)}{\partial t} = k\int_x^\infty x'\left(\frac{x}{x'}\right)^2 \frac{\partial F_1(x',t)}{\partial x'}dx'.$$

Let us transform this equation in terms of the similarity variable η. Using the substitution $F(x,t) = \phi(\eta)$, and using $\eta = x/h(t)$, the left-hand side becomes

$$\frac{\partial F(x,t)}{\partial t} = \frac{d\phi(\eta)}{d\eta}\frac{\partial\eta}{\partial t} = -\frac{d\phi(\eta)}{d\eta}\frac{x(dh/dt)}{h^2} = -\frac{\eta}{h}\frac{d\phi}{d\eta}\frac{dh}{dt}.$$

Treating the right-hand side in a similar fashion:

$$k\int_x^\infty x'\left(\frac{x}{x'}\right)^2 \frac{\partial F_1(x',t)}{\partial x'}dx' = k\int_\eta^\infty \eta' h(t)\left(\frac{\eta}{\eta'}\right)^2 \frac{d\phi(\eta')}{d\eta'}\frac{\partial\eta'}{\partial x'}h(t)d\eta'.$$

Combining and simplifying gives

$$-\frac{\eta}{h^2}\frac{d\phi}{d\eta}\frac{dh}{dt} = k\int_\eta^\infty \eta'\left(\frac{\eta}{\eta'}\right)^2 \frac{d\phi(\eta')}{d\eta'}d\eta'.$$

Now, recall that the purpose of similarity transformation is to reduce the PDE into an ODE, which means the equation after the transformation should be only in terms of η and there should not be any explicit x or t dependance. It can be readily seen that if we set

$$\frac{1}{h^2}\frac{dh}{dt} = \text{constant}$$

we achieve our goal. The value of the constant can be chosen arbitrarily and we can set it to $-k$ so that it cancels with the k in the right-hand side. The negative sign is to stop indefinite growth of h at long times. The similarity variable is defined since $h(t)$ is known from

$$\frac{dh}{dt} = -kh^2$$

and the breakage equation becomes

$$\frac{d\phi}{d\eta} = \int_{\eta}^{\infty} \left(\frac{\eta}{\eta'}\right) \frac{d\phi(\eta')}{d\eta'} d\eta'.$$

Similarity Solution of the Breakage Equation

Defining $\psi(\eta) = d\phi(\eta)/d\eta$ the above equation becomes

$$\psi(\eta) = \int_{\eta}^{\infty} \frac{\eta}{\eta'} \psi(\eta') d\eta'.$$

Differentiating the above equation and applying the Leibnitz rule:

$$\frac{d\psi(\eta)}{d\eta} = -\psi(\eta) + \int_{\eta}^{\infty} \frac{1}{\eta'} \psi(\eta') d\eta'.$$

Differentiating again gives

$$\frac{d^2\psi(\eta)}{d\eta^2} = -\psi'(\eta) - \frac{1}{\eta}\psi(\eta).$$

Rearranging gives

$$\eta \frac{d^2\psi(\eta)}{d\eta^2} + \eta \frac{d\psi(\eta)}{d\eta} + \psi(\eta) = 0,$$

which is a second-order linear ODE with variable coefficients. We will need two boundary conditions for this equation. Finding appropriate boundary conditions, the solution of this equation, and back transformation to number density are left to the reader as an exercise. One important point regarding the similarity solution is that it is usually the solution attained at long times. Because the initial distribution is arbitrary, it takes some time for the distribution to attain the self similar form. For many practical cases, however, the self similar form is reached quickly.

7.3 Numerical Solution of the PBE

A large number of numerical techniques are available to solve PBEs. Discretization methods are a very popular and versatile numerical technique for the solution of PBEs. Finite difference, finite volume, and finite elements are often used, although finite difference has limited applicability for breakage–aggregation problems because of the fine grid spacing requirement. Another useful technique for numerical solution of the PBE is the quadrature method of moments. In this book we shall discuss only a particular type of finite volume discretization technique known as the fixed pivot technique. The MATLAB codes for solving aggregation and breakage problems using this technique are given in sections C.2 and C.3.

7.3.1 Discretization Using Finite Volume

The first step in discretization of the PBE is to subdivide the internal coordinate space into smaller fragments. In most cases of practical importance the particle size space spans many orders of magnitude. For example, for aerosol dynamics, small nanometre size particles exists along with particles that are tens of micrometres in size. Hence, the particle volume spans over 12 orders of magnitude. If a linear grid capable of capturing the small as well as large particles is to be used, we will require 10^{12} nodes. As we will see, we will need to solve a differential equation for each node and hence the computational task will be massive. Moreover, most of these nodes will be located towards the large size where nanometre resolution is not required.

To resolve this issue, a geometric grid is used instead of a linear grid. In a geometric grid, the smallest node is located at a very small size (v_1, say, 10^{-27} m^3) and then every successive node point is multiplied by a ratio (r, say, 1.4) to obtain the next node. Hence, the nth node will be located at $v_1 \times (r)^n$ on the size axis. Therefore, to cover a size range from 10^{-27} m^3 to 10^{-15} m^3, we need only 82 nodes and hence only 80 ODEs need to be solved.

Although it may appear that it should not be a problem to solve a few hundred equations and one can afford a finer grid, repeated calculations using PBE are often performed for optimization. Hence, a small problem size is always helpful. However, the accuracy of the solution must be monitored carefully because a coarse grid is being used at large size. A portion of the geometric grid is shown in Figure 7.3.

In finite volume discretization, the equation is integrated over the finite volume of the bin (as shown in Figure 7.3 as alternate shaded and unshaded regions) and the PBE is reduced to a set of ODEs. The number density distribution is therefore represented by the number of particles on nodes. These nodes are also called 'pivots' [3] and are located at the centre of each bin. To illustrate this method, let us consider the breakage equation:

$$\frac{\partial f_1(v,t)}{\partial t} = -\Gamma(v)f_1(v,t) + \int_v^{\infty} \Gamma(v')f_1(v',t)v(v')P(v|v')dv'.$$

Integrating this equation over the ith bin and denoting $N_i = \int_{v_i}^{v_{i+1}} f_1(v,t)dv$ we get

$$\frac{dN_i}{dt} = -\int_{v_i}^{v_{i+1}} \Gamma(v)f_1(v,t)dv + \int_{v_i}^{v_{i+1}} \int_{v}^{\infty} \Gamma(v')f_1(v',t)\nu(v')P(v|v')dv'dv. \tag{7.4}$$

Now, the left-hand side of the equation is in terms of the number of particles in the ith bin, but the right-hand side is in terms of an unknown number density which cannot be integrated readily to obtain a closed-form set of ODEs in terms of N_i. To resolve this problem, the number density is approximated as a series of Dirac-delta functions on the nodes:

$$f_1(v,t) \approx \sum_{j=1}^{\infty} N_j \delta(v - x_j).$$

Substituting this in the right-hand side of the above equation and using the property of the Dirac-delta function, the first term on the right becomes $\Gamma_i N_i$.

To deal with the second term, we first note that this is the breakage birth term. Hence, it represents the birth of particles on the ith pivot by breakage of particles from larger pivots. To be consistent with daughter distribution, however, a particle on a pivot may form particles whose size does not match that of any of the pivots. Such non-pivot particles must be assigned to the nearest pivots because we are only counting particles on pivots. During this assignment, the mass and number of particles are preserved.

Let us consider the formation of n_v particles of size v which fall between the ith and $(i+1)$th pivots. Also assume that L_i of these particles will be assigned to ith pivot (node) and R_{i+1} of these will be assigned to the $(i+1)$th pivot (see Figure 7.3). These assignments should be such that mass and number are preserved. These two preservations are written as

$$\begin{aligned} L_i + R_{i+1} &= n_v \\ x_i L_i + x_{i+1} R_{i+1} &= v n_v. \end{aligned}$$

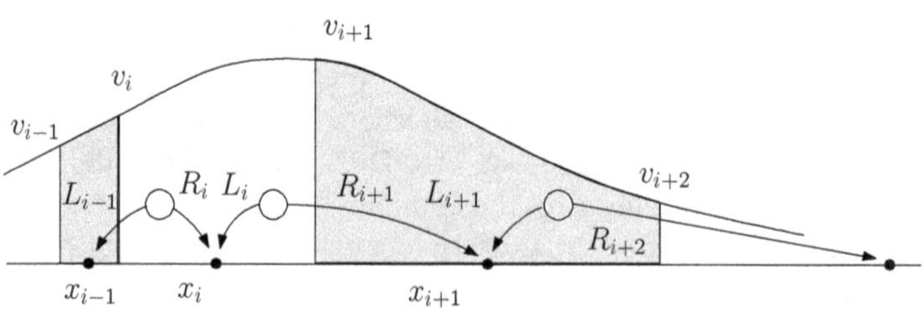

Figure 7.3 Discretization of space using a geometric grid. The bin boundaries and nodes are shown.

Setting $n_v = 1$ (per particle basis) and solving these equations:

$$R_{i+1}(v) = (v - x_i)/(x_{i+1} - x_i)$$
$$L_i(v) = (x_{i+1} - v)/(x_{i+1} - x_i).$$

With these assignment rules, the birth term for the ith pivot will have contributions from the left (with the right fraction) and from the right (with the left fraction) 'inter-pivot' regions. Hence the breakage birth term in eqn 7.4 can be rewritten as

$$B_{bi} = \int_{x_{i-1}}^{x_i} \int_{v}^{\infty} R_i \Gamma(v') f_1(v', t) \nu(v') P(v|v') dv' dv$$
$$+ \int_{x_i}^{x_{i+1}} \int_{v}^{\infty} L_i \Gamma(v') f_1(v', t) \nu(v') P(v|v') dv' dv.$$

Substituting the approximate expression for number density and considering the binary breakage ($\nu(v') = 2$):

$$B_{bi} = 2 \sum_{x_k \geq x_i} N_k \Gamma_k \int_{x_{i-1}}^{x_i} R_i(v) P(v|x_k) dv + 2 \sum_{x_k \geq x_i} N_k \Gamma_k \int_{x_i}^{x_{i+1}} L_i(v) P(v|x_k) dv$$
$$= 2 \sum_{x_k \geq x_i} \Gamma_k N_k \left(\int_{x_{i-1}}^{x_i} R_i(v) P(v|x_k) dv + \int_{x_i}^{x_{i+1}} L_i(v) P(v|x_k) dv \right).$$

The term in parentheses is known as soon as the discretization grid is defined. Denoting this as n_{ik}, the breakage equation becomes

$$\frac{dN_i}{dt} = -\Gamma_i N_i + 2 \sum_{x_k \geq x_i} n_{ik} \Gamma_k N_k.$$

The breakage PBE (the partial integro-differential equation) is reduced to a set of ODEs which can be solved numerically to obtain N_i. To obtain the number density, N_i should be divided by the width of the ith bin. The term involving aggregation can be dealt with in a similar fashion, and nucleation and growth can be included in the numerical scheme. However, for the growth process a high-resolution scheme by Gunawan [4] is more popular. If more than one internal coordinate is used, a multi-dimensional grid is needed [5, 6].

7.4 Exercises

Exercises 7.1 Derive the equation for the zeroth and first moments for the aggregation PBE using the constant aggregation kernel.

Exercises 7.2 Derive the moment of the growth equation for constant and linear growth rates.

Exercises 7.3 Write down the PBE considering nucleation, growth, and breakage. Obtain the equation for the jth moment for homogeneous nucleation, constant growth rate (G), and linear breakage rate ($B = B_0x$).

Exercises 7.4 Solve the breakage PBE for the breakage functions given in section 7.2.3 using the similarity solution. Compare the solution with the numerical solution from the fixed pivot technique.

Bibliography

1 R. McGraw. Description of aerosol dynamics by the quadrature method of moments. *Aerosol Science and Technology, 27*(2):255–265, 1997.
2 D. Ramkrishna. *Population Balances: Theory and Applications to Particulate Systems in Engineering*. Academic Press, 2000.
3 S. Kumar and D. Ramkrishna. On the solution of population balance equations by discretization–I. A fixed pivot technique. *Chemical Engineering Science, 51*(8):1311–1332, 1996.
4 R. Gunawan, I. Fusman, and R. D. Braatz. High resolution algorithms for multidimensional population balance equations. *AIChE Journal, 50*(11):2738–2749, 2004.
5 J. Chakraborty and S. Kumar. A new framework for solution of multidimensional population balance equations. *Chemical Engineering Science, 62*(15):4112–4125, 2007.
6 C. Y. Ma, X. Z. Wang, and K. J. Roberts. Morphological population balance for modeling crystal growth in face directions. *AIChE Journal, 54*(1):209–222, 2008.

8

Kinetic Monte Carlo Simulation

The description of a particle often involves more than one internal coordinate. For many practical problems, such as cell culture or granulation, there can be as many as ten internal coordinates. For such cases, formulation of the population balance model becomes complicated and even the numerical solution of such models is difficult. Kinetic Monte Carlo (KMC) simulation is used for such cases.

In many non-linear systems fluctuation around the mean becomes important. Hence, mean field models like population balances are inadequate for such cases. (For details of stochastic systems and their models, the reader is referred to [1, 2]. For a thorough treatment of Monte Carlo method, the reader is referred to Ref. [3].) KMC simulation should also be used for such systems. In this chapter we will discuss the KMC simulation technique.

KMC simulation starts with a finite (often small) population of particles and lets individual entities in the population evolve according to certain average rates allowing for random scatter around the mean rate. The formulation of the model for KMC simulation is simple but the computation required is often very lengthy. The key feature for KMC simulation is to recognize the randomness of the physical variables and estimate their distribution. We will start with a brief discussion of random variables and random numbers before we discuss the algorithm.

8.1 Random Variables

The value of a random variable can change from one replicate of the experiment to another. Hence it is not possible to specify a single value for a random variable even for a set of tightly specified conditions. Instead, random variables are characterized by a set of values with associated probabilities. In other words, we can specify the probability distribution of random variables but not any single value for them.

Engineering of Submicron Particles: Fundamental Concepts and Models, First Edition. Jayanta Chakraborty.

Companion website: http://booksupport.wiley.com

It is important to note the difference between random variables and deterministic variables. For a deterministic variable, if the circumstances are well defined, the variable has unique value. For example, if a ball is dropped on the surface of the earth from a particular height in vacuum, its free fall velocity after a certain time is unique and hence is a deterministic variable. In contrast, if the same ball falls in the presence of atmospheric air, its velocity becomes a random variable because it is affected by uncontrollable factors like the properties of the surrounding air. The probability distribution of a random variable may be a probability mass function or a probability density function depending on whether the variable is discrete or continuous.

Example 8.1 Specify the probability distribution of the following random variables: (a) the number of heads in a sequence of five independent tosses of a fair coin and (b) the time between two successive calls in a call centre.

Solution: Let X denote the random variable 'the number of heads in a sequence of five independent tosses of a fair coin'. Note that this is a discrete random variable. The values that X can take are 0, 1, 2, 3, 4 and 5. The probability mass function $f(X)$ for this case will be

$$f(X = n) = \frac{5!}{n!(5-n)!}(0.5)^n(1-0.5)^{(5-n)}.$$

The range of values and its probability distribution completes the description of this random variable.

The time between two calls at a call centre is another example of a random variable. Can we specify a finite set of values for the variable as before? The time between two calls can be 2 minutes or 5 minutes but it can also be 2.1345 minutes. It is clear that this variable is a continuous random variable. Let us denote it by Y, which spans in the interval $[0, \infty]$. The probability distribution of this variable is given by the Poisson distribution

$$F(Y = y) = \frac{\lambda^y e^{-\lambda}}{y!},$$

where λ is the average rate of the arrival process. The derivation of the two distribution functions given above is available in Montgomery [4]. □

8.1.1 Uniform Random Numbers

A uniform random number is a random variable with equal probability of its occurrence in a given interval, usually from 0 to 1.0. One of the ways to obtain a set of uniform random numbers is to follow a truly random process such as the falling of raindrops on a surface. For practical usage, however, we need computers to generate uniform random numbers. Hence we need an algorithm for the generation of uniform random numbers.

There is a fundamental fallacy in this statement. If we have an algorithm for producing uniform random numbers, that will no longer be random. Therefore, the uniform random numbers produced by computers using an algorithm (like the Rnd# button on your calculator) are not true random numbers and are termed 'pseudo random numbers'. Such numbers are actually a sequence of numbers with negligible correlation. For practical use, such pseudo random numbers are sufficient.

There is a variety of random number generators ranging from very simple to complex. In the following example, we demonstrate a simple random number generator. However, such a simple generator is not very useful and we need to use more sophisticated random number generators whose code can be obtained from *Numerical Recipes* [5]. Commercial software packages like Matlab also have good random number generators as a built-in function.

Example 8.2 Consider the following random number generator:

$$\zeta_{i+1} = \text{mod}\ (a\zeta_i + c, N),$$

where ζ_i is the sequence of random numbers, a, c, and N are three given constants, and the notation $mod\ (a, b)$ produces the remainder when a is divided by b (e.g. $\text{mod}\ (13, 5) = 3$).

- Generate 30 random numbers using $a = 263$, $c = 71$, $N = 100$, and $\zeta_0 = 79$. Mark any notable feature.
- If the set of parameters is changed to $a = 2631$, $c = 711$, $N = 100$, and $\zeta_0 = 79$, what difference does it make to the random numbers?
- What is the effect of changing ζ_0?

Solution: A simple code can be written to generate the random sequence, which is obtained as:

$$\begin{array}{ccccccccccccccc} 79 & 48 & 95 & 56 & 99 & 8 & 75 & 96 & 19 & 68 & 55 & 36 & 39 & 28 & 35 \\ 76 & 59 & 88 & 15 & 16 & 79 & 48 & 95 & 56 & 99 & 8 & 75 & 96 & 19 & 68 \end{array}$$

A dot plot of these numbers shows that they are not clustered and are seemingly distributed randomly. However, the series repeats after 20 terms. Changing the parameters by an order of magnitude increases the length of the unrepeated sequence to 50. Usually a very large number is used for a and b. Changing the seed (ζ_0) changes the sequence completely. □

8.2 Algorithm for KMC Simulation

There are two types of KMC simulation: event-driven and time-driven. (Recently a number of variants have been published, e.g. see Ref. [6].) In this chapter we shall consider event-driven KMC simulation. In this algorithm the system is advanced in time by administering the events one by one. In time-driven KMC simulation the system is advanced in time and then all the events that occurred during that interval are administered.

To illustrate the idea of event-driven KMC simulation we will consider the process of particle breakage and hence the 'event' is the breakage of a particle. As discussed before, we will consider the rate of breakage to be a random variable distributed around the mean breakage rate. The mean breakage rate will be obtained using the breakage functions.

8.2.1 Specification of the System

In KMC simulation the evolution of a system (e.g. a box containing 100 particles) is followed in time and the first step in this algorithm is to represent the system, as discussed in section 5.2. Since KMC simulation is normally used for multi-dimensional populations, the particle array is often multi-dimensional, as shown below. In this case, the particle volume, the mass of a component, and the location in Cartesian coordinates are tracked.

As the system evolves by occurrence of events, this array must be updated after each of the events. It should be noted that the number of particles in this array changes dynamically due to birth–death processes. For example, each breakage event adds a new member (column) to the array and modifies one column. The attribute of the individual member can also evolve due to processes like growth.

$$\mathbf{A(t)} = \begin{pmatrix} v_1 & v_2 & v_3 & v_4 \\ m_1 & m_2 & m_3 & m_4 \\ x_1 & x_2 & x_3 & x_4 \\ y_1 & y_2 & y_3 & x_4 \\ z_1 & z_2 & z_3 & x_4 \end{pmatrix}$$

8.2.2 Time between Events: Interval of Quiescence

Usually events occur randomly with some known average rate of occurrence [7]. Hence the process may be considered as Poisson's process and the time between events is a continuous random variable with exponential distribution. For KMC simulation, the time between events is called the interval of quiescence (IQ) and its probability density function is given by

$$f(\tau) = \lambda \exp(-\lambda\tau), \tag{8.1}$$

where λ is the mean rate of the process, for example breakage. If multiple events are occurring, λ will be the sum of the rates of all possible events.

Example 8.3 Let us consider the random variable 'time spent by a patient in a dentist's surgery'. The dentist sees on an average 20 patients every day and stays for 4 hours in her surgery. Obtain the expression for the probability that a patient will spend less than x minutes in the surgery. What is the probability that the dentist will attend a patient for 30 minutes or more?

Solution: The average time spent by the dentist per patient is 12 minutes. Hence the average rate at which events occur is (number of events/minute, λ) 1/12. The probability that the event will occur between 0 minutes and x minutes is given by integrating the probability density function within these two limits:

$$F(x) = \int_0^x \lambda \exp(-\lambda\tau)d\tau = 1 - \exp(-\lambda x).$$

Hence, the probability that the event will not occur within the first x minutes is $\exp(-\lambda x)$. Setting $x = 30$, we obtain the probability as 0.08. □

It can be seen that the average rate of occurrence of events (λ) is needed to obtain the probability distribution of the inter-event time, the IQ. This rate is system specific and as discussed before is the sum of the rates of all possible events. For example, if we consider a system with N_1 particles each with exactly the same size x_1, the number of breakages per unit time will be $N_1\Gamma(x_1)$, which should be used as λ. If we have an additional N_2 particles each with size exactly x_2, the rate of breakage of such particles will be $N_2\Gamma(x_2)$ and λ will be $N_1\Gamma(x_1) + N_2\Gamma(x_2)$.

If the particle size is distributed with a discrete distribution $N_i(x_i)$, the total breakage rate will be $\sum N_i\Gamma(x_i)$. This rule applies even if the number of particles in each size class is only one. In that case the total rate of breakage will be given by $\sum \Gamma(x_i)$. In our case, since the system is represented by individual particles rather than a histogram, the last relation should be used for computing the rate of events, λ.

Example 8.4 Calculate the total rate of breakage for a box containing four particles with volumes 100, 250, 280, and 250 μm. The breakage function $\Gamma(x) = 0.01x.\ \text{s}^{-1}$.

Solution:

$$\begin{aligned}\lambda &= 0.01 \times 100 + 0.01 \times 250 + 0.01 \times 280 + 0.01 \times 250 \\ &= 8.8 \text{ breakages/second}\end{aligned}$$

□

8.2.3 Sampling a Distribution

Once λ is known, the probability distribution of IQ is known. However, in the simulation we will need a sample from this distribution in order to advance

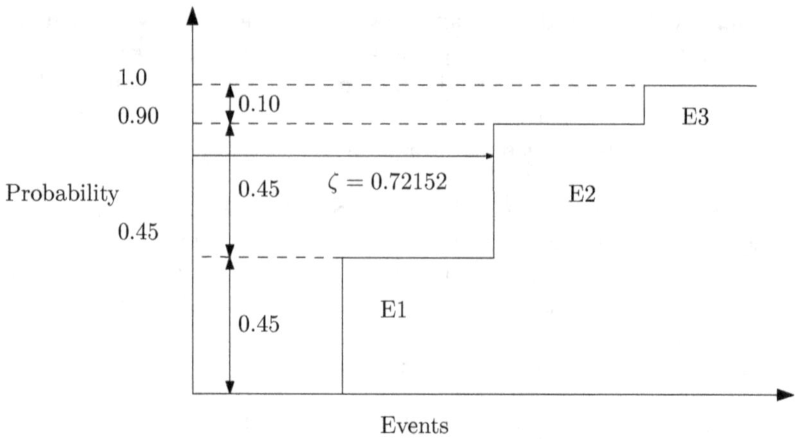

Figure 8.1 Selection of events for a discrete probability distribution.

the system clock and administer the event. How do we generate a sample so that it statistically represents the distribution? If the sample represents the distribution statistically, large numbers of such samples must regenerate the distribution itself. The method used to generate such random samples is first demonstrated using a discrete distribution and then extended for a continuous distribution.

Let us consider a discrete distribution of three events, E_1, E_2, and E_3, with probabilities of occurrence 0.45, 0.45, and 0.1, respectively. The normalized cumulative plot of this distribution is shown in Figure 8.1. Since it is normalized, the y axis will always span from 0 to 1 and is divided among the three events in proportion to their probabilities. If we generate a large number of *uniform* random numbers, 45% of those will fall in the region corresponding to E_1, 45% will fall in the region corresponding to E_2, and 10% will fall in the region corresponding to E_3. Hence such a pool of samples will satisfy the probability distribution of events. Any one of these can be taken as a statistically meaningful sample. To our advantage, we take the first one. If the first random number generated is 0.72152, Event 2 has to be selected.

A similar procedure is used for continuous distribution. Like the discrete case, first we have to obtain the cumulative probability distribution ($F(\tau)$) similar to that shown in Figure 8.2. Then we have to generate the uniform random number (ζ), locate it on the cumulative probability axis, and obtain the corresponding sample value of the random variable from x axis, as shown in Figure 8.2. In a computer procedure, the equation $F(\tau) = \zeta$ is solved for τ for a given ζ. The Matlab code for sampling a continuous distribution is given in section C.5.

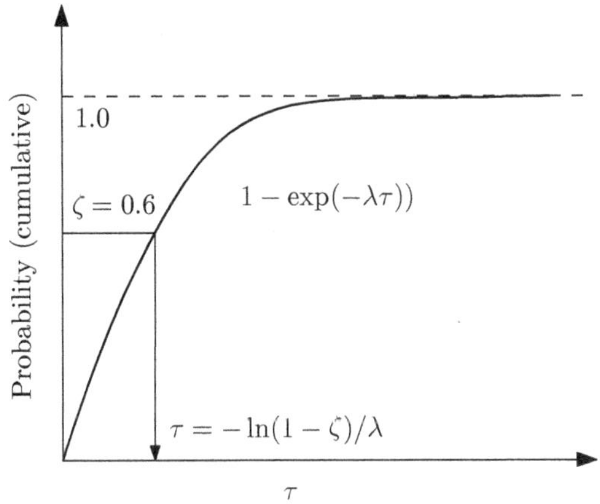

Figure 8.2 Selection of IQ: sampling from a continuous distribution.

Example 8.5 Create a random password using all letters and special characters.

Solution: In this password, all letters and special characters are equally probable. If we take 26 uppercase letters and 10 special characters, we have a total of 36 characters. Hence the range of uniform random numbers (which is 1.0) is to be divided into 36 *equal* segments. The width of each domain will be $\approx$ 0.027778. Let us order the letters and special characters as [A....Z ! @ # .... *()] and we are looking for a four-letter password. A sequence of four random numbers 0.244, 0.945, 0.123, and 0.030 will give IYEB as the password. □

Note that only one sample needs to be drawn in order to advance the system clock. Once the event is administered and the time is advanced, the PSD changes. This change in PSD changes the average rate (λ) of the process and hence the probability distribution of the random variable τ. Hence the next sample is actually drawn from a different distribution. For this reason more than one IQ sample cannot be drawn at the same time.

8.2.4 Events and their Registration

Once the IQ has been selected, the system clock is advanced by this time and the event (breakage in this case) is executed. Now the question is which particle will break? Since it is a stochastic system, the particle with the higher breakage rate will have a higher probability of breakage. Do we know anything about relative probability of breakage of these particles? Since we have knowledge of the breakage rate, which is often a function of size, we may derive the relative probability from the breakage rate.

If the breakage rate is given by $\Gamma(x) = 3.5x^{1.5}$, the probability of breakage of individual particles can be taken proportional to this breakage rate, i.e.

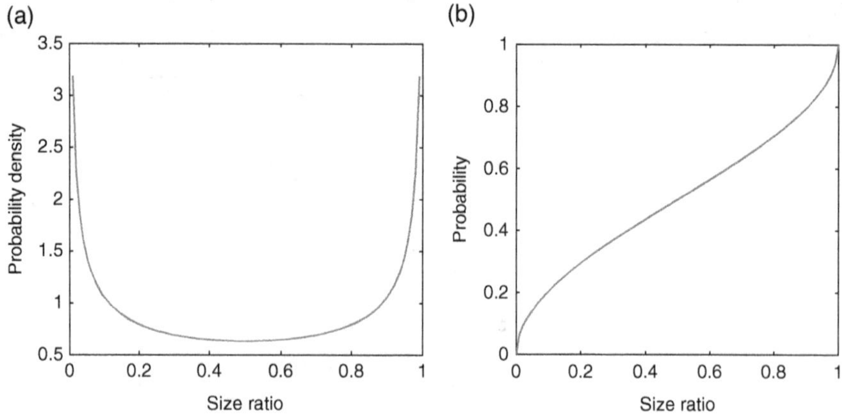

Figure 8.3 Daughter distribution functions: (a) probability density and (b) cumulative probability.

$P_i \propto \Gamma_i$. Hence, if we consider a system containing three particles of size 4, 9, and 2, the breakage probability of these three particles will be 0.2115, 0.7137, and 0.0747, respectively. To select the particle for breakage, we generate a random number and follow the same procedure as shown in Figure 8.1. For example, if the random number is 0.317, particle 2 will break.

Once the particle to be broken has been selected, it is broken according to the physical process governing its breakage. For example, if it is a binary breakage with two equal daughter particles, two particles of equal size are created and the original particle is removed. If it is a binary breakage but the daughter particles are produced according to a certain probability density function, proper probabilistic sampling is needed.

The same machinery for sampling as discussed above can be used to generate samples for daughter particles. For example, let us consider the probability density function for the daughter size distribution shown in Figure 8.3a, which represents the attrition process. To sample this distribution, the probability density function is first converted into a cumulative probability plot, as shown in Figure 8.3b. A random number is then generated and the sample size ratio is obtained. For example, if the random number is 0.194, the size ratio will be 0.11. Hence, if the mother particle has size 9, it will break into two fragments (again, assuming binary breakage) of sizes 8.11 and 0.89.

Finally, the particle array must be updated. An useful convention is to insert the first daughter drop into the place of the mother drop and add the other drop(s) at the end of the array. Hence, for the case above, the updated particle array will be [4 8.11 2 0.89]. A simple code for implementing the KMC algorithm for breakage is given in section C.6.

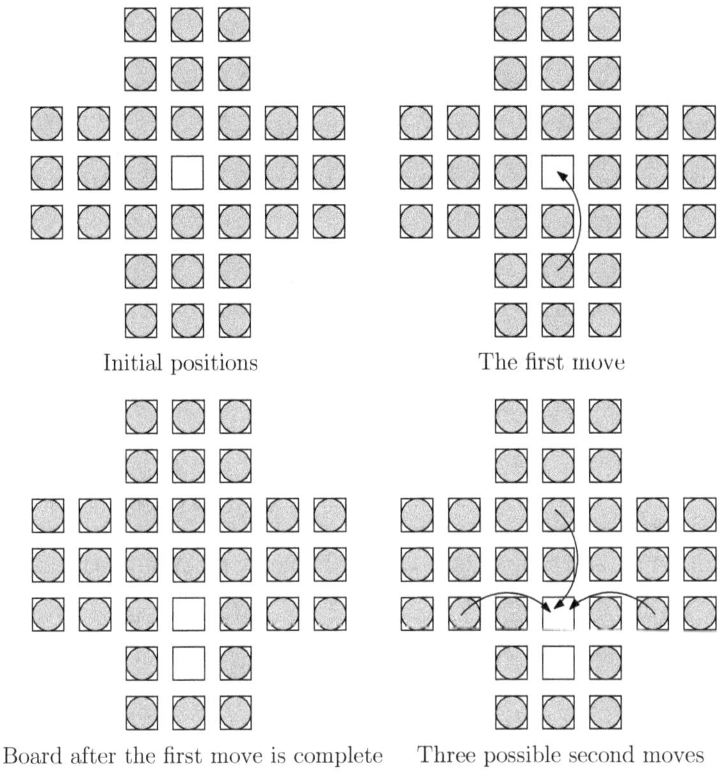

Figure 8.4 Progress of a peg solitaire game.

Example 8.6 A board game called peg solitaire/solo noble/brainvita is popular in many countries. The board at various stages in the game is shown in Figure 8.4. It has total of 33 positions and all except the central one contain a piece. In each move the player has to move one of the pieces in such a way that it jumps over another piece. Only x–y movements are allowed, diagonal movements are not permitted. Jumping over two or more pieces is also not allowed. The piece that has been jumped over is removed from the board. The goal is to reduce the number of pieces on the board. When no permissible move is possible, the game ends and the number of piece on the board is counted. The minimum number is, of course, one and with some practice it is possible to end the game with just one piece remaining. Develop a code to simulate this game and observe how the number of pieces changes as you change the random seed. The computer will choose the next move randomly from all permissible moves.

Solution: The code for the solution of this problem is provided in section C.7. □

8.3 Exercises

Exercises 8.1 A square dart board of unit dimensions has a circle inscribed. Darts are thrown randomly to this board. Using kinetic Monte Carlo simulation show that the fraction of darts that falls inside the circle approaches a value of $\pi/4$ as the number of simulations increases. Consider no dart falls outside the square box.

Exercises 8.2 Consider a system consisting of three particles of volumes 2, 5, and 7. These particles undergo breakage. The breakage rate is proportional to their volumes with the proportionality constant 1. Find the state of the system after 10 events using pen and paper calculation. Use the same seed for random number generation and check your simulation with the code given in section C.6.

Exercises 8.3 Consider the simultaneous breakage and aggregation of particles. Select any breakage and aggregation kernel you prefer. Simulate this case using kinetic Monte Carlo simulation. Tune the kernel so that a steady particle size distribution is achieved.

Bibliography

1 C. W. Gardiner. *Handbook of Stochastic Methods for Physics, Chemistry and the Natural Sciences.* Springer-Verlag, 1983.
2 B. V. Gnedenko. *The Theory of Probability.* Mir Publishers, 1969.
3 D. P. Kroese, T. Taimre, and Z. I. Botev. *Handbook of Monte Carlo Methods.* John Wiley & Sons, 2011.
4 D. C. Montgomery, G. C. Runger, and N. F. Hubele. *Engineering Statistics,* 5th edition. John Wiley & Sons, 2011.
5 W. H. Press, S. A. Teukolsky, W. T. Vetterling, and B. P. Flannery. *Numerical Recipes 3rd Edition: The Art of Scientific Computing*, 2nd edition. Cambridge University Press, 2002.
6 Y. Lin, K. Lee, and T. Matsoukas. Solution of the population balance equation using constant-number Monte Carlo. *Chemical Engineering Science, 57*(12):2241–2252, 2002.
7 D. Ramkrishna. *Population Balances: Theory and Applications to Particulate Systems in Engineering*. Academic Press, 2000.

Appendix A

Mathematical Topics

A.1 Geometry of a Heterogeneous Drop

Consider a drop on a surface, as shown in Figure A.1. The contact angle of this drop is θ and the differential element is chosen at an angle of ϕ. The volume of the differential element marked in the figure can be written as

$$dV_D = \pi(R\sin\phi)^2(R\sin\phi d\phi)$$

$$\begin{aligned}
\therefore \quad V_D &= \int_0^\theta \pi(R\sin\phi)^2(R\sin\phi d\phi) \\
&= \pi R^3 \int_0^\theta \sin^3\phi d\phi \\
&= \pi R^3 \int_0^\theta \frac{1}{4}(3\sin\phi - \sin 3\phi)d\phi \\
&= \frac{3\pi R^3}{4}\int_0^\theta \sin\phi d\phi - \frac{\pi R^3}{4}\int_0^\theta \sin 3\phi d\phi \\
&= \frac{3\pi R^3}{4}[-\cos\phi]_0^\theta + \frac{\pi R^3}{12}[\cos 3\phi]_0^\theta \\
&= \frac{3\pi R^3}{4}[1-\cos\theta] + \frac{\pi R^3}{12}[\cos 3\theta - 1] \\
&= \frac{3\pi R^3}{4}[1-\cos\theta] + \frac{\pi R^3}{12}[4\cos^3\theta - 3\cos\theta - 1] \\
&= \frac{3\pi R^3}{4} - \frac{3\pi R^3}{4}\cos\theta + \frac{\pi R^3}{3}\cos^3\theta - \frac{\pi R^3}{4}\cos\theta - \frac{\pi R^3}{12} \\
&= \frac{2\pi R^3}{3} - \pi R^3\cos\theta + \frac{\pi R^3}{3}\cos^3\theta \\
&= \frac{\pi R^3}{3}[2 - 3\cos\theta + \cos^3\theta] \\
&= \frac{\pi R^3}{3}[2 - 2\cos\theta - \cos\theta + \cos^2\theta - \cos^2\theta + \cos^3\theta]
\end{aligned}$$

Engineering of Submicron Particles: Fundamental Concepts and Models, First Edition. Jayanta Chakraborty.

Companion website: http://booksupport.wiley.com

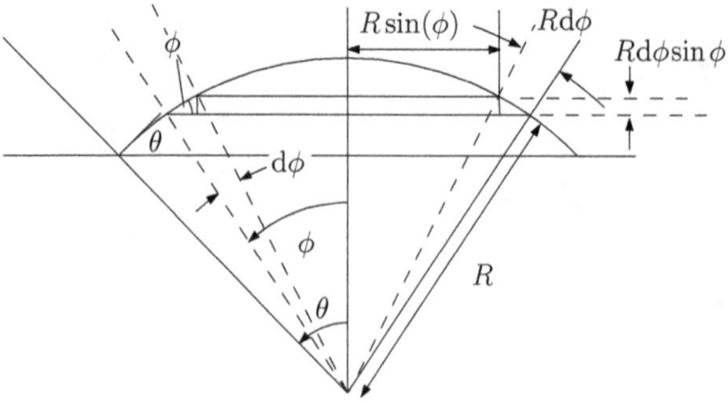

Figure A.1 Geometry of a drop on a surface.

$$= \frac{\pi R^3}{3}[2(1-\cos\theta) - \cos\theta(1-\cos\theta) - \cos^2\theta(1-\cos\theta)]$$
$$= \frac{\pi R^3}{3}(1-\cos\theta)(2-\cos\theta-\cos^2\theta)$$
$$= \frac{\pi R^3}{3}(1-\cos\theta)(2-2\cos\theta+\cos\theta-\cos^2\theta)$$
$$= \frac{\pi R^3}{3}(1-\cos\theta)^2(2+\cos\theta).$$

Similarly,

$$A_C = \int_0^\theta 2\pi R\sin\phi R d\phi$$
$$= 2\pi R^2[-\cos\phi]_0^\theta$$
$$= 2\pi R^2[1-\cos\theta]$$

and $A_b = \pi R^2 \sin^2\theta$.

A.2 Young's Equation

Consider the case shown in Figure A.2 where a drop is residing on a liquid surface and stays at mechanical equilibrium. Hence, a force balance in the horizontal plane (dotted line) yields

$$\gamma_{ov}\cos\theta_1 + \gamma_{ow}\cos\theta_2 = \gamma_{wv}\cos\theta_3.$$

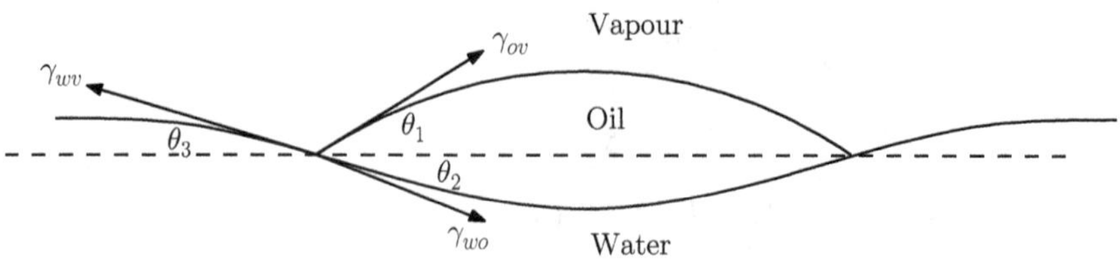

Figure A.2 Oil lens on a water surface: Young's equation.

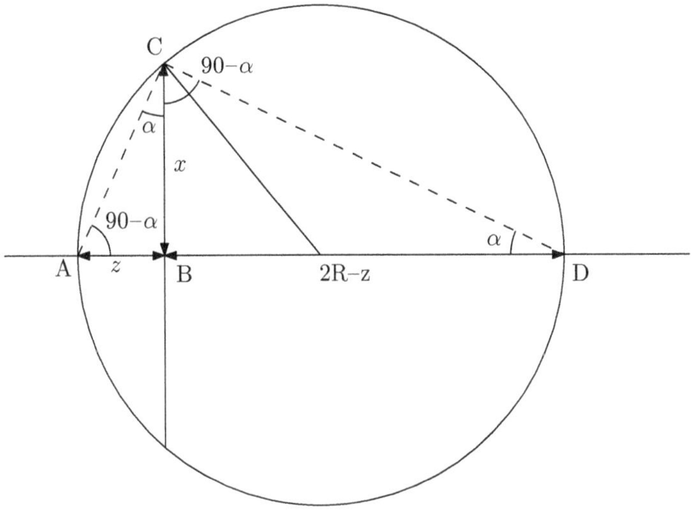

Figure A.3 Chord theorem.

For a solid surface, $\theta_2 = \theta_3 = 0$, $o = l$, and $w = s$. Hence,

$$\gamma_{lv}\cos\theta_1 + \gamma_{ls} = \gamma_{sv}.$$

This is known as Young's equation.

A.3 Chord Theorem

From Figure A.3 it can be seen that the triangles ABC and BCD are similar. Hence,

$$\frac{x}{z} = \frac{2R - z}{x},$$

which gives the chord theorem as $x^2 = z(2R - z)$.

A.4 Jacobian of Variable Transformation in a Multiple Integral [1]

Let us first consider a change of variable for the integration of a simple function, $\phi(x) = x$:

$$\int_0^1 \phi(x)dx = \int_0^1 xdx = 0.5.$$

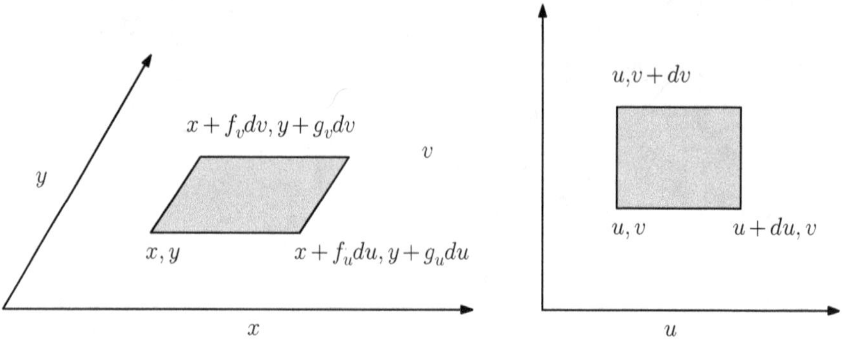

Figure A.4 Transformation of coordinates.

Now, we define another variable $u = x/2$ such that $x = f(u) \equiv 2u$, which signifies shrinking of the axis. In terms of this new variable, the integral may be written as

$$\int_0^{0.5} \phi(2u)(2du) = \int_0^{0.5} (2u)(2du) = 4\int_0^{0.5} udu = 0.5.$$

Note that both the function and the differential 'element' needed a 'transformation'. This transformation can be viewed as shrinking the axis to half its length, due to which the differential element also shrinks. The transformation is trivial for this case and may be written as

$$\int_0^1 \phi(x)dx = \int_0^{0.5} \phi[f(u)]\left(\frac{df}{du}\right)du = \int_0^{0.5} \phi[f(u)]f_u du.$$

The term f_u is called the *Jacobian* of the transformation.

A more systematic approach is needed for the two-dimensional case. Let us consider a variable change from (x, y) to (u, v) where the functional relations between these variables are given by $x = f(u, v)$ and $y = g(u, v)$. The transformation is shown in Figure A.4.

It should be noted that under this transformation the straight line remains straight and hence we need to locate the corners to evaluate the elementary area. The location of the right bottom corner point in the (x, y) system may be written as $(f(u + du, v), g(u + du, v))$. Expanding this in Taylor series and neglecting higher order terms:

$$f(u + du, v) = f(u, v) + \frac{\partial f}{\partial u}du = x + f_u du$$

$$g(u + du, v) = g(u, v) + \frac{\partial g}{\partial u}du = y + g_u du.$$

The other vertices can also be found in a similar fashion.

Now let us express the elementary area in the (x, y) system in terms of (u, v). Since the elementary area in the (x, y) system is a parallelogram, its area is given by

$$\begin{vmatrix} 1 & 1 & 1 \\ x & x+f_u du & x+f_v dv \\ y & y+g_u du & y+g_v dv \end{vmatrix}.$$

This determinant is readily reduced to

$$\begin{vmatrix} f_u du & f_v dv \\ g_u du & g_v dv \end{vmatrix} = dudv \begin{vmatrix} f_u & f_v \\ g_u & g_v \end{vmatrix} = \frac{\partial(x,y)}{\partial(u,v)} dudv$$

$$\therefore \int\int_R \phi(x,y)dxdy = \int\int_{R^*} \phi(f(u,v), g(u,v)) \frac{\partial(x,y)}{\partial(u,v)} dudv,$$

where $\partial(x, y)/\partial(u, v)$ is the two-dimensional Jacobian.

A.5 Method of Characteristics [2]

Let us consider an unknown function $u(x, y)$ given by a partial differential equation of the form

$$au_x + bu_y = f(x, y),$$

where $f(x, y)$ is a known function. In general we can express $u(x, y)$ as

$$u_x dx + u_y dy = du.$$

Comparing these two equations:

$$\frac{dx}{a} = \frac{dy}{b} = \frac{du}{f(x,y)}.$$

This equation is called the auxiliary equation. From the first pair,

$$bx - ay = c,$$

where c is a constant independent of x and y. This equation also yields

$$x = \frac{ay+c}{b}.$$

From the second pair,

$$\frac{dy}{b} = \frac{du}{f\left(\frac{ay+c}{b}, y\right)}$$

$$\text{or} \quad du = \frac{f\left(\frac{ay+c}{b}, y\right)}{b} dy$$

$$\text{or} \quad u = \int \frac{f\left(\frac{ay+c}{b}, y\right)}{b} dy + c_1 \quad (c_1 \text{is a function of } c)$$

$$\text{or} \quad u = \int \frac{f\left(\frac{ay+c}{b}, y\right)}{b} dy + \phi(c)$$

$$\text{or} \quad u = \int \frac{f\left(\frac{ay+c}{b}, y\right)}{b} dy + \phi(bx - ay).$$

The function ϕ and the constant c can be evaluated from the initial and boundary conditions.

Example A.1 Solve

$$2u_x - 3u_y = \cos x \quad u = 1 \quad @ \quad y = 0.$$

Solution: First, let us write down the auxiliary equation:

$$\frac{dx}{2} = \frac{dy}{-3} = \frac{du}{\cos x}.$$

From the first pair we get $3x + 2y = c$. From the second pair,

$$\frac{dx}{2} = \frac{du}{\cos x},$$

which gives

$$u = \frac{1}{2} \sin x + \phi(3x + 2y).$$

Applying the boundary condition,

$$1 = \frac{1}{2} \sin x + \phi(3x),$$

which gives

$$\phi(\zeta) = 1 - \frac{1}{2} \sin(\zeta/3).$$

Hence, the solution is

$$u = \frac{1}{2} \sin x + 1 - \frac{1}{2} \sin\left(\frac{3x + 2y}{3}\right).$$

□

This method works even when a and b are functions of x, as demonstrated next.

Example A.2 Solve

$$u_x + e^x u_y = y \quad u(0, y) = 1 + y.$$

Solution:

$$\text{Auxiliary equation:} \quad \frac{dx}{1} = \frac{dy}{e^x} = \frac{du}{y}.$$

From the first pair we get $y = e^x + c$. We choose the first and third terms to form the second pair:

$$(e^x + c)dx = du$$
$$u = e^x + cx + c_1(c)$$
$$u = e^x + (y - e^x)x + \phi(y - e^x).$$

Using the boundary condition:

$$1 + y = 1 + \phi(y - 1)$$
$$\phi(\zeta) = \zeta + 1.$$

Hence the solution is

$$u = e^x + (y - e^x)x + y - e^x + 1 = 1 + y + x(y - e^x).$$

□

Bibliography

1 E. Kreyszig. *Advanced Engineering Mathematics*, 8th edition. John Wiley & Sons, 1999.
2 P. K. Kythe, M. R. Schäferkotter, and P. Puri. *Partial Differential Equations and Mathematica*. CRC Press, 1997.

Appendix B

Solution of Selected Problems

B.1 General Problem Solving Strategy

It is advisable that the reader practise the following four-step strategy for solving a problem. This part of this text has been taken, often verbatim, from *How to Solve It* by G. Polya [1].

1) **Understand the problem:** Understand the system. Pay extra attention to the details. What are the unknowns? What are the data? What are the conditions? Is it possible to satisfy the conditions? Are the conditions sufficient to determine the unknowns? Or insufficient? Or redundant? Or contradictory? Ask many hows and whys. Draw a diagram, mark all known quantities in the diagram. Mark all unknowns in the diagram with a separate colour or with any other clear visual distinction. Write down all conditions.
2) **Devise a plan:** Find the connection between the data and the unknown. You may be obliged to consider an auxiliary problem if some immediate connection cannot be found. You have to obtain eventually a plan for the solution. List all questions that should be answered. Consider the following aspects:
   - Have you seen it before? Have you seen the same problem in a slightly different form?
   - Do you know a related problem or any theory that could be useful?

   These may provide valuable clues.
3) **Carry out your plan:**
   - Write down the precise quantities you require.
   - Represent the system by a simpler one, keeping the key features.
   - Write down all balance equations (mass balance, energy balance, momentum balance, population balance).
   - Write down all constitutive equations (relation between fluid velocity and viscosity, growth rates as a function of supersaturation etc.)
   - Check whether they form a closed set of equations that can be solved.
   - Make additional assumptions and/or write additional equations if the set is not closed. Assumptions should be justified, verifiable, and consistent.

   Solve the set of equations. Check the solution for correctness.

Engineering of Submicron Particles: Fundamental Concepts and Models, First Edition. Jayanta Chakraborty.

Companion website: http://booksupport.wiley.com

4) **Look back:** Can you check the physical reality of the results? Can you check for limiting cases? Are all arguments sound? Can you reach any of the results using a different approach? Can you use this method for any other situation?

B.2 Solutions of Selected Problems

Answer to Exercise 1.1 Let us assume that all the particles are spherical. Now, we need to find the equilibrium solute concentration for particles of different sizes. For the larger particles:

$$\ln \frac{c(r = 0.5\mu\text{m})}{c_\infty} = \frac{2\gamma\bar{v}''}{rRT} = \frac{2 \times 500 \times 10^{-3} \times \frac{150}{3} \times 10^{-6}}{0.5 \times 10^{-6} \times 8.314 \times 300} = 0.04.$$

which gives

$$\frac{c(r = 0.5\mu\text{m})}{c_\infty} = 1.04.$$

For the smaller particles:

$$\ln \frac{c(r = 0.1\mu\text{m})}{c_\infty} = \frac{2\gamma\bar{v}''}{rRT} = \frac{2 \times 500 \times 10^{-3} \times \frac{150}{3} \times 10^{-6}}{0.1 \times 10^{-6} \times 8.314 \times 300} = 0.2,$$

which gives

$$\frac{c(r = 0.1\mu\text{m})}{c_\infty} = 1.22.$$

It can be seen that for the larger particles the solution is slightly unsaturated. Hence, they will start to dissolve. But because the driving force is small, the dissolution will be very slow. However, the degree of unsaturation is much higher for the smaller particles. Hence, these particle will dissolve very fast.

Upon complete dissolution of the smaller particles, the concentration will go up and will become 225 g/l (neglecting the dissolution of large particles). Hence, the solution will have a concentration $1.125c_\infty$, but any concentration above $1.04c_\infty$ will be sufficient to grow the larger particles. Hence, although they were dissolving slowly at the beginning, they will start to grow after the fine particles dissolve.

If only 8 g of the smaller particles are added, the new supersaturation ratio after dissolution will be 1.04, which is the saturation concentration of the larger particle. This will stop the dissolution of larger particles, but no growth of larger particles will occur.

Answer to Exercise 1.2 Because the bubble is very small, gravity effects are not important and the bubble try to form a spherical cap on the tip. Initially a very small volume of gas is present in the bubble and hence its radius of curvature is very large. Gradually, the volume of gas increases and the radius of curvature decreases and approaches that of the capillary tube. After it reaches a hemispherical shape, its radius again increases until it detaches from the capillary tube. Note that the bubble cannot maintain a constant angle at the tip unlike a bubble on a flat surface because its ends are pinned to the edge of the tube. Because the radius of the sphere goes through a minimum, the pressure drop goes through a maximum.

Answer to Exercise 1.6 First let us write down the Kelvin equation for this system. The two conditions for the system are $v'dp' = v''dp''$ and $dp' - dp'' = d(2\gamma/r)$. Note that the second expression is different for this case compared to the usual 'suspended drop is a vapour' case. Hence, the Kelvin equation becomes

$$\ln(p'/p_0) = -\frac{2\gamma v''}{rRT}.$$

Hence, for the 8 nm cavity,

$$\ln(p'/p_0) = \frac{-2 \times 75 \times 10^{-3} \times 18 \times 10^{-6}}{8 \times 10^{-9} \times 8.314 \times 373} = -0.11,$$

giving $p'/p_0 = 0.89$. Hence, at a saturation ratio of 0.9, the cavity will gradually fill. However, when it reaches the broader part, which is 17 nm, it will stop because the saturation ratio corresponding to this dimension is 0.949.

When the saturation ratio is raised to 0.95, however, this part will also fill up and it will also fill up the entire cavity, including the narrower neck at the top. When the supersaturation ratio is brought back to 0.9, the top neck cannot evaporate and the cavity remains filled.

Answer to Exercise 1.7 The rate of nucleation is given by

$$\dot{N} = k_{n1}[\ln(p'/p_0)]^2 \exp\left(-\frac{k_{n2}}{[\ln(p'/p_0)]^2}\right).$$

Now the pre-exponential factor can be written as

$$\begin{aligned} k_{n1}[\ln(p'/p_0)]^2 &= \frac{3}{2}\frac{D}{d^5}\sqrt{\frac{(kT)^3}{\pi\psi^3}}\left[\frac{\Delta H_v}{R}\left(\frac{1}{T}-\frac{1}{T_0}\right)\right]^2 \\ &= \frac{3}{2}\frac{D}{d^5}\sqrt{\frac{k^3}{\pi\psi^3}}\left(\frac{\Delta H_v}{R}\right)^2 T^{3/2}\left(\frac{1}{T}-\frac{1}{T_0}\right)^2 \\ &= k_1 T^{3/2}\left(\frac{1}{T}-\frac{1}{T_0}\right)^2. \end{aligned}$$

Hence, the rate of nucleation becomes

$$\dot{N} = k_1 T^{3/2}\left(\frac{1}{T} - \frac{1}{T_0}\right)^2 \exp\left(-\frac{4\left[\frac{\psi}{3kT}\right]^3}{\left[\frac{\Delta H_v}{R}\left(\frac{1}{T} - \frac{1}{T_0}\right)\right]^2}\right)$$

$$= k_1 T^{3/2}\left(\frac{1}{T} - \frac{1}{T_0}\right)^2 \exp\left(-\frac{k_2}{T^3\left(\frac{1}{T} - \frac{1}{T_0}\right)^2}\right).$$

The constants k_1 and k_2 are related to physical variables. Although such variables are also a function of temperature, their dependance is much less. This expression can be differentiated and set to zero to obtain the T at which the rate of nucleation is maximum. Defining $T/T_0 = y$, the above expression becomes

$$\dot{N} = k_1' \frac{(1-y)^2}{\sqrt{y}} \exp\left(-\frac{k_2'}{y(1-y)^2}\right),$$

which finally gives

$$\frac{(5 - y_{max})(1 - y_{max})}{2y_{max}(1 + y_{max})} = k_2'.$$

Answer to Exercise 4.1 The Matlab code for calculating the potential is given in section C.1. Using this code, the κ is calculated to be 3.6×10^8 and the corresponding Debye length is 2.78 nm. The constants $c_1 = 228.38$ and $c_2 = 1.62 \times 10^{-7}$. The plot of Φ_{net}/kT versus distance is shown in Figure B.1. It can be seen that the maximum potential is $30kT$ and it occurs at 2 nm separation distance. The second derivative of $\frac{\Phi_{net}}{kT}$ can be written as

$$\frac{d^2}{dr^2}\left(\frac{\Phi_{net}}{kT}\right) = c_1\kappa^2 \exp(-\kappa r) - \frac{2c_2}{r^3}.$$

At the point of maximum potential:

$$\left.\frac{d^2}{dr^2}\left(\frac{\Phi_{net}}{kT}\right)\right|_{r=r_{max}} = c_1\kappa^2 \exp(-\kappa r_{max}) - \frac{2c_2}{r_{max}^3} = -2.62 \times 10^{19}.$$

Hence, $p = 3.6 \times 10^9$, which gives

$$w = \frac{\sqrt{\pi}\exp(30)}{4 \times 1 \times 10^{-7} \times 3.6 \times 10^9} = 1.3 \times 10^9.$$

The slow coagulation constant k_r(slow) is therefore given by

$$k_r(\text{slow}) = \frac{k_r(\text{fast})}{w} = \frac{1.5 \times 10^{-17}}{1.3 \times 10^9} = 1.15 \times 10^{-26}.$$

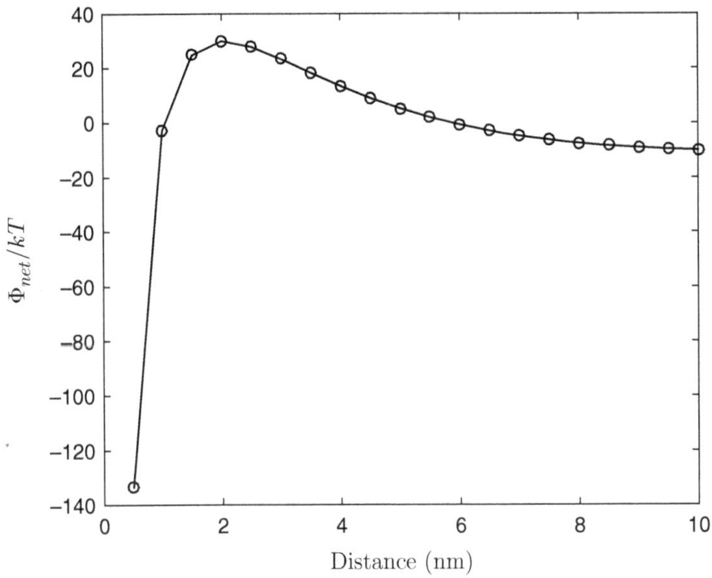

Figure B.1 Potential versus distance.

The half life therefore is

$$t_{1/2} = \frac{1}{c_0 \times k_r(slow)} = 10^{10}s = 317 \text{ years}.$$

Answer to Exercise 4.2 The interaction energy between two rings as shown in Figure 4.10 can be written as

$$d(\Phi_R/kT) = 2\pi h dh \frac{64\rho_0\gamma_0^2}{\kappa} \exp(-\kappa z).$$

It is clear from the geometry that there exists a relation between h and z and hence dh and dz. Hence, such a relation is required for integration of the potential given above. Using chord theorem,

$$\frac{h}{2R - \frac{z-s}{2}} = \frac{\frac{z-s}{2}}{h}.$$

Denoting $\frac{z-s}{2} = k$, the above expression can be written as

$$k^2 - 2Rk + h^2 = 0,$$

which can be solved to

$$k = R - \sqrt{R^2 - h^2}.$$

The unrealistic root has been removed (since $k < R$) in the above expression. Now substituting for k,

$$\frac{z-s}{2} = R - \sqrt{R^2 - h^2}.$$

Differentiating the above expression,

$$\frac{dz}{2} = -\frac{-2hdh}{2\sqrt{R^2 - h^2}}$$

or

$$dz = \frac{2hdh}{\sqrt{R^2 - h^2}}.$$

Now, the interaction energy can be written as

$$d\Phi_R/kT = \pi\sqrt{R^2 - h^2}\frac{64\rho_0\gamma_0^2}{\kappa}\exp(-\kappa z)dz.$$

Hence,

$$\Phi_R/kT = \int_s^{2R} \pi\sqrt{R^2 - h^2}\frac{64\rho_0\gamma_0^2}{\kappa}\exp(-\kappa z)dz$$

$$\Phi_R/kT = \int_s^{2R} \pi R\sqrt{1 - (h/R)^2}\frac{64\rho_0\gamma_0^2}{\kappa}\exp(-\kappa z)dz.$$

If the major contribution to the interaction energy comes from the location near the approach point, h/R can be considered to be small and can be neglected. Because potential drops significantly at larger distance, no significant error is introduced by extending the limit of the integration to ∞. With these considerations,

$$\Phi_R/kT = \int_s^{\infty} \pi R\frac{64\rho_0\gamma_0^2}{\kappa}\exp(-\kappa z)dz$$

$$= \frac{64\pi R\rho_0\gamma_0^2}{\kappa^2}\exp(-\kappa s).$$

Answer to Exercise 5.3 For unknown growth rates of particles of different sizes we can look at the situation in following ways.

Case 1: $G(d) \propto d$

When $d_1 > d_2$, particles with higher diameter grow much faster than those with d_2, so smaller particles can never exceed the higher particle size. Hence $d_2 > d_1$ is never possible for this case.

Case 2: $G(d) \propto \frac{1}{d}$

Particles with lower diameter will grow faster than those with higher diameter, but as they grow more, rate of growth will be decreased, thus after sometime the smaller particles with diameter d_2 may approach the size of bigger particle.

Thereafter they will grow at the same rate. Hence, in this situation also, $d_2 > d_1$ is not possible.

Case 3: G is independent of size
In this case, particles of smaller size cannot grow faster than those of higher size, so at any later time d_2 cannot be more than d_1.

Answer to Exercise 5.6 The total mass of crystal per unit volume of the crystallizer is given by

$$M_T = \int_0^\infty \rho\alpha L^3 n(L) dL,$$

where $\alpha = \pi/6$. For MSMPR crystallizer,

$$n(L) = n_0 \exp\left(\frac{-L}{G\tau}\right).$$

Substituting the above relation:

$$\begin{aligned}
M_T &= \rho\alpha n_0 \int_0^\infty L^3 \exp\left(\frac{-L}{G\tau}\right) dL \\
&= \rho\alpha n_0 \left[L^3 \int_0^\infty \exp\left(\frac{-L}{G\tau}\right) dL + G\tau \int_0^\infty 3L^2 \exp\left(\frac{-L}{G\tau}\right) dL\right] \\
&= 3\rho\alpha n_0 G\tau \int_0^\infty L^2 \exp\left(\frac{-L}{G\tau}\right) dL \text{ (The first term vanishes when evaluated)} \\
&= 6\rho\alpha n_0 (G\tau)^2 \int_0^\infty L \exp\left(\frac{-L}{G\tau}\right) dL \\
&= 6\rho\alpha n_0 (G\tau)^3 \int_0^\infty \exp\left(\frac{-L}{G\tau}\right) dL \\
&= 6\rho\alpha n_0 (G\tau)^4.
\end{aligned}$$

Denoting M_L the mass of crystal up to a size L:

$$\begin{aligned}
M_L &= \int_0^L \rho\alpha\xi^3 n(\xi) d\xi \\
&= \rho\alpha n_0 \int_0^L \xi^3 \exp\left(\frac{-\xi}{G\tau}\right) d\xi \\
&= \rho\alpha n_0 \left[\xi^3 I + 3G\tau \int_0^L \xi^2 \exp\left(\frac{-\xi}{G\tau}\right) d\xi\right] \\
&= \rho\alpha n_0 \left[\xi^3 I + 3G\tau \left(\xi^2 I + 2G\tau \int_0^L \xi \exp\left(\frac{-\xi}{G\tau}\right) d\xi\right)\right] \\
&= \rho\alpha n_0 [\xi^3 I + 3G\tau(\xi^2 I + 2G\tau\{\xi I + G\tau I\})]_0^L.
\end{aligned}$$

Substituting

$$I = \int_0^L \exp\left(\frac{-\xi}{G\tau}\right) d\xi = -G\tau \exp(-\xi/G\tau)\Big|_0^L = -G\tau\left[\exp(-L/G\tau) + 1\right]$$

$$M_L = \rho\alpha n_0 \left[\{L^3 + 3L^2 G\tau + 6L(G\tau)^2 + 6(G\tau)^3\}(-G\tau)\exp\left(\frac{-L}{G\tau}\right) + 6(G\tau)^4\right].$$

Denoting $x = L/(G\tau)$,

$$M_L = \rho\alpha n_0 (G\tau)^4[-(x^3 + 3x^2 + 6x + 6)e^{-x} + 6]$$

or

$$\frac{M(x)}{M(T)} = \left[1 - \left\{1 + x + \frac{x^2}{2} + \frac{x^3}{6}\right\} e^{-x}\right].$$

$M(x)$ is the cumulative of the mass density distribution. Hence, the mass density distribution, denoted by $m(x)$, is obtained by differentiating the above expression:

$$m(x) = \frac{d}{dx}\left[\frac{M(x)}{M(t)}\right] = -\left[1 + x + \frac{x^2}{2}\right] e^{-x} + \left[1 + x + \frac{x^2}{2} + \frac{x^3}{6}\right] e^{-x} = \frac{x^3}{6} e^{-x}.$$

Answer to Exercise 5.10 In this case, the number density function is in two dimensions. Hence, it looks like a hill. The internal coordinates are cell age τ and cell mass m, respectively. Hence, the balance equation can be written as

$$\begin{aligned}\frac{d}{dt} f_1(\tau, m, t)d\tau dm = {} & \dot{m}(m, c(t), \tau) f_1(\tau, m, t)d\tau|_m \\ & - \dot{m}(m, c(t), \tau) f_1(\tau, m, t)d\tau|_{m+dm} + f_1(\tau, m, t)dm|_\tau \\ & - f_1(\tau, m, t)dm|_{\tau+d\tau},\end{aligned}$$

which gives

$$\frac{\partial}{\partial t} f_1(\tau, m, t) + \frac{\partial}{\partial \tau} f_1(\tau, m, t) + \frac{\partial}{\partial m}\dot{m}(m, c(t), \tau) f_1(\tau, m, t) = 0. \quad \text{(B.1)}$$

Answer to Exercise 7.1 The aggregation equation is (for $a_0 = 1$)

$$\begin{aligned}\frac{\partial f(x,t)}{\partial t} = {} & \frac{1}{2}\int_{x'=0}^{x'=x} f(x - x', t) f(x', t)dx' \\ & - \int_{x'=0}^{x'=\infty} f(x, t) f(x', t)dx'.\end{aligned} \quad \text{(B.2)}$$

After moment operation:

$$\begin{aligned}\int_0^\infty x^j dx \frac{\partial f(x,t)}{\partial t} = {} & \int_0^\infty x^j dx \frac{1}{2}\int_{x'=0}^{x'=x} f(x - x', t) f(x', t)dx' \\ & - \int_0^\infty x^j dx \int_{x'=0}^{x'=\infty} f(x, t) f(x', t)dx',\end{aligned} \quad \text{(B.3)}$$

which becomes

$$\frac{d\mu_j}{dt} = \frac{1}{2}\int_{x=0}^{x=\infty} dx \int_{x'=0}^{x'=x} dx' x'^j f(x-x',t) f(x',t) - \mu_0\mu_j.$$

The first term on the right becomes (after change in limit)

$$\frac{1}{2}\int_{x'=0}^{x'=\infty} dx' \int_{x=x'}^{x=\infty} dx x'^j f(x-x',t) f(x',t).$$

Substituting $x - x' = u$ and setting $dx = du$, we have

$$\frac{1}{2}\int_{x'=0}^{x'=\infty} dx' \int_{u=o}^{u=\infty} du (u+x')^j f(u,t) f(x',t).$$

For the general case, the term $(u + x')^j$ can be expanded using binomial theorem. For $j = 2$ this becomes

$$\frac{1}{2}\int_{x'=0}^{x'=\infty} dx' \int_{u=o}^{u=\infty} du (u^2 + x'^2 + 2ux') f(u,t) f(x',t) = \mu_2\mu_0 + \mu_1\mu_1,$$

which gives

$$\frac{d\mu_2}{dt} = \mu_2\mu_0 + \mu_1\mu_1 - \mu_0\mu_2 = \mu_1^2.$$

Bibliography

1 G. Polya. *How to Solve It: A New Aspect of Mathematical Method*, 2nd Edition. Princeton University Press, 1971.

Appendix C

Codes

C.1 Distance-Dependant Potential

```
clear all
clc
close all

%========= Universal Constants =============
e=1.603e-19; % Electronic charge
k = 1.38e-23; % Boltzmann constant
epsi_0 = 8.85e-12; % Permittivity of vacuum

% === Parameters for dispersing medium =====
A = 8e-20; % Hamaker's constant
T=298; % Absolute temperature
epsi_r = 78.5; % Relative permittivity
z=1; % Valency of positive ions
v(1)=1; % Valency of ion type 1
v(2)=-1; % Valency of ion type 2
c0= 12e-3; % Conc of electrolyte in (Molar)
n0=c0*1e3*6.023e23; % Conc in no/m^3
% ============================================

% ====== Particle Parameters ================
R=1e-7; %Particle Size in m
psi_0 = 50e-3; % Surface potential in Volts

% ==== Required Functions ===================
gamma_0=(exp((z*e*psi_0)/(2*k*T))-1)/...
    (exp((z*e*psi_0)/(2*k*T))+1);
epsi = epsi_r*epsi_0;

% ========= Calculate kappa^2 ===============
kappa2 = 0;
```

Engineering of Submicron Particles: Fundamental Concepts and Models, First Edition. Jayanta Chakraborty.

Companion website: http://booksupport.wiley.com

```
for i=1:2
kappa2= kappa2 +(v(i)^2)*e^2*n0/(epsi*k*T);
end
kappa = sqrt(kappa2)

% ========= Calculate constants =============
c1=(64*pi*R*n0*gamma_0^2)/(kappa^2)
c2 = (A*R)/(12*k*T)

% ====== Calculate potential ================
r=0.5e-9;
% At various distances starting from 0.5 nm
for i=1:250
att = c2/r; % Attractive potential
rep = c1*exp(-kappa*r); % Repulsive potential
net = rep-att;
pot(i,:)=[r net];
r=r+0.5e-9;
end

% ======  Plot the potential   =============
semilogx(pot(:,1)/1e-9, pot(:,2),'-*')
xlabel('Distance (nm)')
ylabel('Potential (T)')
```

C.2 Solution of Breakage PBE

```
% Code for solution of breakage population
% balance equation for binary breakage
% using Fixed Pivot Technique.
% Breakage Frequency: \Gamma(v)=v^2
% Daughter Distribution function:
% \beta(v,v') = 2/v'
% To use this code the user required to change
% in the user handle section. If the kernels
% change, the analytical solution should be
% modified for fresh benchmarking. This step
% may be skipped by commenting the benchmarking
% section. Send bug reports to
% jayanta@che.iitkgp.ac.in

% Disclaimer: This code has been written for a
% beginner. Hence, all error checks, modularity
% and generality have been sacrificed for the
% sake of simplicity.
```

```
function [] = breakage_code()
close all;
clear all;
clc;

%============== User handle starts =============
%% User defined parameters for  grid
v_min = 1e-5; % Minimum  size/volume
v_max = 1.0;  % Maximum  size/volume
r = 1.5;      % Geometric ratio
m_v = ceil(log(v_max/v_min)/log(r))+2; % Number
% of volume nodes. Because the largest pivot will
% be smaller than the largest volume specified, we
% increase the number of volume nodes by 2.

m = m_v -1; % Number of pivots is 1 less than the
% number of volume nodes.
%%


%% Provide initial conditions here.
disp('Setting up initial distribution.')
% N0(i) is the initial number of particles on pivot i.
N0 = zeros(m,1);
N0(end) = 1.0;
% Initial monodispersed condition on largest pivot
tspan = [0 4000];  %time span for integration
t_enq = 3000; % The time at which the number density
% will be plotted.
%%

%%
disp(['Number of pivots:',num2str(m)])

% Breakage rate function
    function Gamma = Gammafun(x)
        Gamma = (x).^2; % Power law breakage
    end
%
% Daughter Distribution function
    function Beta = Betafun(v,v_prime)
        Beta = 2/v_prime; % Uniform distribution
    end
%%
% ===============User handles ends ===============
```

```
%% Geometric grid generation
disp('Creating the geometric grid.')
% Creating the bin boundaries.
vi=zeros(m_v,1);
vi(1)=v_min;
for i=2:m_v
    vi(i) = vi(i-1)*r;
end
x=zeros(m,1); % Pivots
width = zeros(m,1);
% Generating the pivots and bin width
for i=1:m
    x(i)=(vi(i+1)+vi(i))/2;
    width(i) = vi(i+1)-vi(i);
end
Gam = Gammafun(x); % Breakage rates for Pivots
%%

%==================================================
disp('Generating n(i,k).')
%% Generating the nik matrix
% n(i,k)is a matrix that provides the breakage
% birth to population at ith pivot due to the
% breakage of a particle on kth pivot. For details,
% see Kumar and Ramkrishna, Chem. Eng. Sci, 1996
% pp-1311-1332. See Eq. 29 of the ref.

n=zeros(m,m);

% The second integral is zero for the first pivot.
 i = 1;
 for k=2:m
 Fun_1 = @(v)((x(i+1)-v)*Betafun(v,x(k))...
     /(x(i+1)-x(i)));
 n(i,k)= integral(Fun_1,x(i),x(i+1));
 end

 for i=2:m
 k = i;
 % The first integral is zero for the diagonal terms.
 Fun_2 = @(v)((v-x(i-1))*Betafun(v,x(k))...
     /(x(i)-x(i-1)));
 n(i,k) = integral(Fun_2,x(i-1),x(i));

% Terms above the main diagonal
 for k=i+1:m
```

```
 Fun_1 = @(v)((x(i+1)-v)*Betafun(v,x(k))...
     /(x(i+1)-x(i)));
 Fun_2 = @(v)((v-x(i-1))*Betafun(v,x(k))...
     /(x(i)-x(i-1)));
 n(i,k)=integral(Fun_1,x(i),x(i+1)) + ...
    integral(Fun_2,x(i-1),x(i));
end
end
%% ================================================


disp('Solving the ODEs')
tic
%% solver for the discretized equations
[t,N] = ode45(@pbe,tspan,N0);
%%
disp('ODE Sol completed successfully!')
toc

%================================================
%% Specification of discretized equations (ODEs)
function dN_dt = pbe(t,N)
dN_dt = zeros(m,1);
for i = 1:m
% Compute the birth term
birth = 0;
for k = i:m
birth = birth + n(i,k)*Gam(k)*N(k);
end
% Complete discretized PBE
dN_dt(i) = birth - Gam(i)*N(i);
end

end % ODE definitions ends here.
%================================================


%% ============== Data presentation =============
% Computing and plotting the zeroth moment.
save('Numden.mat','N','x','t');
disp('Computing and plotting zeroth moment.')
figure(1) % Plotting scaled zeroth moment vs time
Ntot = sum(N,2);
plot(t,Ntot/Ntot(1),'s')
title('Normarized Zeroth Moment')
xlabel('Time')
```

```
ylabel('N(t)/N(0)')
hold on

% Computing and plotting number distribution.
% at t = t_enq.
ta = find(t>t_enq,1,'first'); % Find the time
% that is closest to the enq point.
disp(['Plotting the number distribution at t = '...
    ,num2str(t(ta))])
figure(2)
semilogx(x,N(ta,:),'s')
title('Number distribution')
xlabel('Particle size')
ylabel('Number')
hold on
% %%
% %%=================================================
%
%
%%===Comparison with analytical solution =========
disp('Plotting analytical solution')
% Benchmarking for power law binary breakage and
% uniform daughter distribution. Analytical solution
% is taken from Ziff and McGrady, J. Phys. A, 18
% (1985)3027-37 (eqn. 25)

 N_ana = zeros(m,1);
 tau = t(ta);
for i = 1:m
 N_ana(i) = integral(@(x)exp(-tau*x.^2),vi(i),vi(i+1));
end
 N_ana = N_ana*2*tau*x(m);
 semilogx(x,N_ana)

 end
```

C.3 Solution of Aggregation PBE

```
% Code for solution of aggregation population
% balance equation using Fixed Pivot Technique
% Constant aggregation kernel : Q(v,v') = a0

% Disclaimer: This code has been written for a
% beginner. Hence, all error checks, modularity
% and generality have been sacrificed for the
```

```
% sake of simplicity. Bugs can be reported at
% jayanta@che.iitkgp.ac.in

function [] = aggregation_code()
clear all;
close all;
clc;

% ======== User handle starts ==================
%% Initializing
v_max = 100.0;      % Maximum size/volume
v_min =1e-5;        % Minimum size/volume
r = 2;              % Geometric ratio
v0 = 0.01;          % Initial average size
N0 = 1;             % Initial total number
tspan = [0 2];      % time span for integration

%% Generating discretized grid
disp('Creating the geometric grid..')
m = ceil(log(v_max/v_min)/log(r)); % No pivots
mv = m + 1;        % Number of volume boundaries

vi =  zeros(mv,1);
vi(1)=v_min;
for i=2:mv
    vi(i)=vi(i-1)*r;
end
x=zeros(m,1); % Creating pivots
w=zeros(m,1); % Bin width
for i = 1:m
    x(i) = (vi(i)+vi(i+1))/2;
    w(i) = vi(i+1) - vi(i);
end
%%

%% Setting up initial condition.
% Exponential initial condition has been used.
% n(x,0)=(N0/v)exp(-(v/v0). This expression
% has been integrated from vi(i) to vi(i+1)
% to obtain the initial number of particles
% on ith pivot.
% NOTE: The analytical solution used for
% benchmarking is for this initial condition
% only.
disp('Setting-up initial conditions...')
```

```
Ni0 = zeros(m,1);
for i = 1:m
Ni0(i) = N0*(exp(-vi(i)/v0)-exp(-vi(i+1)/v0));
end

%%

%% Aggregation Kernel
% Constant Kernel
a0 = 1.0;
%% ======== User handle Ends ==================


%% ======== \eta_ijk calculation ==============
% n(i,j,k) is contribution to ith pivot when
% particles on jth and kth pivot aggregate.
% For details see Kumar and Ramkrishna 1996
disp('Generating \eta_ijk..')
eta = zeros(m,m,m);

for j = 1:m
for k = j:m
v = x(j) + x(k);
for i = 1:m-1
    if(v>x(i) && v<= x(i+1))
    eta(i+1,j,k) = (v-x(i))/(x(i+1)-x(i));
    eta(i,j,k)= (x(i+1)-v)/(x(i+1)-x(i));
    break;
    end
end

end
end
%% ===========================================


%% solver
disp('Solving the ODE..')
tic
[t,N] = ode45(@pbe,tspan,Ni0);
toc

%%

%% ========= Discretized equations =============
function dN_dt = pbe(t,N)
dN_dt = zeros(m,1);
```

```
dgnl = eye(m); %This works as Kronecker delta

for i = 1:m

birth = 0;
for j = 1:m
for k = j:m
birth = birth + ...
    (1-0.5*dgnl(j,k))*eta(i,j,k)*N(j)*N(k);
end
end

dN_dt(i) = - N(i)*sum(N)+ birth;
end

end
%% ==============================================

%% Presentation of data
% Plotting and comparing zeroth moment. Closed
% form analytical expression is available for
% zeroth moment for constant kernel.
nt = size(t,1);

figure(1)
disp('Comparing zeroth moment with analytical.')
Ntol = sum(N,2);
loglog(t,Ntol,'s')
xlabel('time')
ylabel('N(t)/N(0)')
hold on

Ntot_ana = zeros(nt,1);
for i=1:nt
Ntot_ana(i) = 1/(1/N0+a0*t(i)/2);
end
loglog(t,Ntot_ana)
legend('Numerical','Analytical')
title('Comparison with analytical zeroth moment')
%

%% Plotting and comparing number density
figure(2)
% Convirt the Ni to density
numden=zeros(nt,m); % Numerical numberdensity
for i=1:nt
    for j=1:m
```

```
        numden(i,j) = N(i,j)/w(j);
        end
end
% Plot numerical number density
loglog(x,numden(nt,:),'*')
axis([1e-4 1e1 1e-1 1e3])
hold on

% Calculate the analytical density.
% See Chakraborty and Kumar, Chemical
% Engineering Science 2007 for the
% expression and reference for the same.
anaden=zeros(nt,m);
for i =1:nt
    tau = a0*N0*t(i);
for j=1:m
    v=x(j);
    sumn = 0;
for k=1:12
sumn = sumn + ...
((tau/(tau+2))^(k-1))*((v/v0)^(k-1))...
/factorial(k-1);
end
anaden(i,j)=(4*N0/((tau+2)^2))*...
    (1/v0)*exp(-v/v0)*sumn;

end
end
% Plot the initial condition.
loglog(x,anaden(1,:))
% Plot the analytical number density
loglog(x,anaden(nt,:))
legend('Numerical','Initial','Analytical')
title('Comparison with analytical number density')
end
```

C.4 Sampling of a Discrete Distribution

```
clear all
close all
clc
% A particle size distribution from
% sieve analysis is given.
% Average size on sieve:
x=[100 200 300 400 500 600 700];
```

```
% Number of particles on each sieve.
n=[30 100 900 700 300 90 10];
nsamp=10; % Number of samples needed.

samp=zeros(nsamp,1);
nsieve=size(x,2);
nsum=sum(n);
n=cumsum(n)/nsum; % Normalized cumulative
% Random number for sampling.
zeta=rand([nsamp,1]);


for j=1:nsamp % Generate samples
% Search the location of zeta
for i=1:nsieve
if(zeta(j)<n(i))
    samp(j)=x(i);
    break
end
end
end
```

C.5 Sampling of a Continuous Distribution

```
function psd = sampling_cont(mu,sigma,nsample)
% Sampling from a normally distributed population.
% Population mean is \mu and variance \sigma^2.
% Generate random number for sampling;
zeta=rand([nsample,1]); % Random number for sampling
sample=zeros(nsample,1); % Initialize sample array


for i=1:nsample
xold=mu; % give the mean as guess value
err=1.0; % Initiate the error

while err>1e-4 % Use newton-Raphson to solve
% 'normcdf' is the cumulative distribution function
% and its derivative is given by 'normpdf'
xnew=xold-(normcdf(xold,mu,sigma)-zeta(i))/normpdf...
    (xold,mu,sigma);
err=abs(xnew-xold)/xold;
xold=xnew;
end
```

```
sample(i)=xnew;  % Pick the sample.
end

psd=sample;
```

C.6 Simulation of Breakage Using KMC

```
% Kinetic Monte Carlo simulation for drop breakage
clear all
close all
clc

%%============== User handle starts ============
% pds is the initial particle size distribution.
% psd=[1 3 8  9  7  2]; % One can start with a simple
% Initial PSD. However, in the following we used the
% initial PSD generated using a normal distribution.

mu = 500;
sigma = 100;
nsample = 50;
% Initial PSD has mean size as 500 micrometre,sigma as
% 100 micrometre and 50 samples are drawn.
psd=sampling_cont(mu,sigma,nsample);

time=0.0;                % set initial time
event=50; % Number of breakage events we want to see
zeta_t=rand(event,1); % Generate random number for IQ
zeta_p=rand(event,1); % and for particle breakage
% Breakage Functions

% Power lab breakage with Gamma(x)  = a x^b
a=3.45;
b=1.5;

% Binary breakage into two equal daughter particles.

%%
%Creating the skeleton for the simulation movie
axis([0 2*mu 0 nsample]) % XY Range
fig1=figure(1);                    % Capture frame
winsize = get(fig1,'Position'); % Set window size
winsize(1:2) = [0 0];
A=moviein(event+1,fig1,winsize);% Movie array
x=2*mu*[0.1 0.2 0.3 0.4 0.5 0.6 0.7 0.8 0.9];
```

```
% Creating the x axis
hist(psd,x); % plots the pds in Figure-1.
A(:,1)=getframe(fig1,winsize); % Save it as frame
xlabel('Particle size')
ylabel('Frequency')
%% ================= User handle ends ==============

%% =============== Simulation Starts ===============
for j=1:event
% Administer events one by one
n=size(psd,1);                  % Number of particles
Gamma=a*psd.^b;                 % Breakage rate vector
lambda=sum(Gamma);              % Total breakage rate
tgap = -log(zeta_t(j))/lambda;  % IQ
time=time + tgap;               % Advance the clock

Gamma_cumul = cumsum(Gamma)/lambda;
% Making the normalizer cumulative of breakage
% rate so that we can select the particle to
% break depending on the relative breakage rate.

brkwhich=0;          % Which one to break?
% Particles with higher breakage rate has higher
% probability to be selected for breakage.
intbegin=0;    % Divide into intervals:
%interval begins
intend=Gamma_cumul(1);     %interval ends
for i=1:n                  % Test each particle
if zeta_p(j)>intbegin && zeta_p(j)<intend
    %If the random number falls in its interval
    brkwhich=i;    % ith particle breaks
    break                  % exit from the loop
end
intbegin=intend;        % if not, continue search.
intend=Gamma_cumul(i+1);
end


psd(brkwhich)=psd(brkwhich)/2;
% break the particle into two
psd(n+1)=psd(brkwhich);
% add the extra particle

hist(psd,x) % Histogram is plotted in Figure-1
axis([0 2*mu 0 nsample])
xlabel('Particle size')
```

```
ylabel('Frequency')
A(:,j)=getframe(fig1,winsize);
% Frame is added to the movie
end

movie(fig1,A,1,10,winsize) % Re-play the movie
```

C.7 Simulation of Brainvita Game

```
% Code for simulating Brainvita game.
clear all
close all
clc
% Along with the 7*7 matrix, two additional
% spaces are added in all four sides. These
% positions are not used (their value set to
% zero) but their presence simplifies the
% algorithm considerably. 0 means illegal
% positions, 1 means filled position and
% 2 means vacant legal position

pos=ones(11,11);
% If the value is zero, no piece can be
% placed in those locations. These values
% cannot be altered.
pos(1:4,1:4)=0;
pos(1:4,8:11)=0;
pos(8:11,1:4)=0;
pos(8:11,8:11)=0;
pos(1:2,5:7)=0;
pos(5:7,1:2)=0;
pos(5:7,10:11)=0;
pos(10:11,5:7)=0;
% At the beginning the centre is vacant.
pos(6,6) = 2; % 2 means empty.

nmove = 10; % Execute 10 moves


for move=1:nmove

pos % Display the board position.

% Identify all empty positions
n_empty=0;
for j=1:11
```

```
for k=1:11
if(pos(j,k)==2)
   n_empty=n_empty+1;
   pos_emp(n_empty,:)=[j k];
end
end
end

% Select empty position for move
% All empty positions are equally likely.
zeta_pos=rand(1,1);
for j = 1:n_empty
    if(zeta_pos <j/n_empty)
    pos_sel = pos_emp(j,:);
    break
    end
end

% Select the move direction and check if
% the move is possible.
zeta_dir=rand(1,1);

if (zeta_dir<0.25) % Northward movement
    if(pos(pos_sel(1)-1,pos_sel(2))==1 ...
            && pos(pos_sel(1)-2,pos_sel(2))==1)
     % A move is possible if the position at
     % north and north-north both are filled.
     pos(pos_sel(1),pos_sel(2)) = 1;
     pos(pos_sel(1)-1,pos_sel(2))= 2;
     pos(pos_sel(1)-2,pos_sel(2))= 2;
     % Complete the move.
    end
elseif (zeta_dir <0.50) % Move to south.
     if(pos(pos_sel(1)+1,pos_sel(2))==1 ...
             && pos(pos_sel(1)+2,pos_sel(2))==1)
     pos(pos_sel(1),pos_sel(2)) = 1;
     pos(pos_sel(1)+1,pos_sel(2))= 2;
     pos(pos_sel(1)+2,pos_sel(2))= 2;
    end
elseif (zeta_dir <0.75) % Move ot east.
    if(pos(pos_sel(1),pos_sel(2)+1)==1 ...
            && pos(pos_sel(1),pos_sel(2)+2)==1)
     pos(pos_sel(1),pos_sel(2)) = 1;
     pos(pos_sel(1),pos_sel(2)+1)= 2;
     pos(pos_sel(1),pos_sel(2)+2)= 2;
    end
```

```
else % Move to west
   if(pos(pos_sel(1),pos_sel(2)-1)==1 ...
           && pos(pos_sel(1),pos_sel(2)-2)==1)
     pos(pos_sel(1),pos_sel(2)) = 1;
     pos(pos_sel(1),pos_sel(2)-1)= 2;
     pos(pos_sel(1),pos_sel(2)-2)= 2;
   end
end

end
```

Appendix D

Experimental Demonstration

$CuSO_4$ is one of the many salts that show very good change in solubility with temperature. The solubility of $CuSO_4$ in water at 100 °C is 75.4 g anhydrous salt/100 g water and it drops to 14.3 g anhydrous salt/100 g water at 0 °C [1]. Hence, a good amount of supersaturation can be created if a saturated solution of $CuSO_4$ at 100 °C is cooled to 25 °C.

To visualize nucleation growth and creation of a population of $CuSO_4$ crystals similar to that in a batch crystallizer, saturated $CuSO_4$ solution at 80 °C was made in a beaker. The water was heated on a hot plate and the salt was gradually added to it until some salt remained at the bottom. Then a small amount of water was added so that the residual salt dissolved and the solution became clear. To check this, some feature through the opposite wall of the glass vessel must be clearly visible and no moving debris should be present. A bright light such as a table lamp should be used, as shown in the video experiment (The video of this process is available at http://booksupport.wiley.com).

The heating can then be turned off or the hot plate may be replaced by another with only stirring. The nucleation-growth starts in about 10 minutes. After the first observation of particles, a large number of particles appear within a few seconds. The particles forming at the beginning are flake-like but they grow to hundreds of microns within the first few seconds because of the high supersaturation. If the stirrer is stopped the particles settle immediately. If the solution is kept overnight, large crystals grow, testifying that growth continues for a prolonged period.

Crystal growth can also be observed using the slurry produced in the process. A microscope with 10X magnification is sufficient to observe the growth process. A solution of $CuSO_4$ saturated at room temperature can be added to the existing crystals to observe their growth. In this case supersaturation

Engineering of Submicron Particles: Fundamental Concepts and Models, First Edition. Jayanta Chakraborty.

Companion website: http://booksupport.wiley.com

is created by evaporation of the water under ambient conditions. A video of crystal growth is also available at http://booksupport.wiley.com.

Bibliography

1 H. Robert and D. W. Green. *Perry's Chemical Engineers' Handbook*, 7th edition. McGraw-Hill, 1997.

Index

Engineering of Submicron Particles: Fundamental Concepts and Models, First Edition. Jayanta Chakraborty.

Companion website: http://booksupport.wiley.com

www.ingramcontent.com/pod-product-compliance
Lightning Source LLC
LaVergne TN
LVHW011521200926
848484LV00001B/101